RIVERSIDE COMMUNITY COLLEGE
1916

P9-CJF-952

THE SOLUTION TO POLLUTION IN THE WORKPLACE

THE SOLUTION TO POLLUTION IN THE WORKPLACE

Laurence Sombke
Terry M. Robertson Elliot M. Kaplan

MasterMedia Limited
NEW YORK

Library of Congress Cataloging-in-Publication Data

Sombke, Laurence.
The solution to pollution in the workplace/Laurence Sombke.
p. cm.
ISBN 0-942361-25-3
1. Pollution—Economic aspects. 2. Environmental protection—Economic aspects. 3. Industrial management—Environmental aspects. 4. Industry—Social aspects. 5. Work environment. I. Title.
HD69.P6S65 1991
363.7'05765—dc20 90-48519
CIP

Designed by Stanley S. Drate/Folio Graphics Co. Inc.

Manufactured in Canada

10 9 8 7 6 5 4 3 2 1

Contents

Foreword

I work in the petroleum industry and consider myself an environmentalist. That may not make a lot of sense to some people who believe that the use of petroleum products and a clean environment are mutually exclusive, but it is true.

"Big Oil" has had a bad reputation with consumers in the past, mainly because of oil spills from tankers and the burning of fossil fuels, which has been linked to air pollution.

Good news! There is a new generation of engineers and executives working in the oil business who *do* care about the environment and who *are* beginning to make their voices heard.

That's why I contributed to this book, *The Solution to Pollution in the Workplace*. I think it is the duty of every citizen and every company to protect the environment for ourselves and for future generations. One person can make a difference! People in business, even a small business, can have a positive impact on a cleaner environment.

For example, at STRATCO, Inc., the company of which I am president and owner, we have an in-house office recycling program. We recycle office paper, cans, bottles and cardboard. In addition, we encourage our associates to recycle at home and bring their recycled material to work, where we have it disposed of properly.

Another reason I was interested in this book was to give America and American businesses both a pat on the back and a little push. I recently traveled overseas and was shocked to see the amount of pollution there. I saw rivers tainted with waste oil, with no attempts being made to clean it up. Raw sewage and industrial waste were being dumped untreated into waterways. These sorts of environmental violations are rare now in the United States because of the hard work and efforts people, governments and businesses expended to clean up.

But there is more work to be done. Individual businesses can start by cleaning up our own backyards. How? Read this book. It is full of information to give your business practical methods of pollution prevention. Set up programs in the workplace. Get your employees involved. Show some leadership. If individuals can do it, so can business! If small business can do it, so can big business! If we all do our part, we can have a cleaner environment for everyone.

—S. Diane Graham
President and CEO,
STRATCO, Inc.

Preface

Now that the hoopla and media blitz of the twentieth anniversary of Earth Day are over, it is time for all of us to roll up our sleeves and get down to the business of cleaning up the environment. And we do mean business, because the title and the focus of this book is *The Solution to Pollution in the Workplace.* Laurence Sombke's first book was simply *The Solution to Pollution,* and that was designed to give tips, techniques and strategies about what individuals can do to clean up their own backyard.

Well, business has a backyard, too, and we are very encouraged to see that businesses, both large and small, are beginning genuinely to care about what goes on in that backyard.

Environmental activists have struggled and labored for the past twenty years to bring environmental cleanup and stewardship up to the level where it is today. Now, it is our belief that fundamental, long-lasting environmental cleanup will take place only if business takes the handoff, if business will accept the baton and run its leg of the environmental relay.

This is a golden moment for business for two reasons. One, they have an opportunity to use the organized, get-it-done strategy of American entrepreneurial capitalism to make this world a cleaner and safer place to live. Second, environmentalism is good for business. You have the chance to cut costs, increase your profits and actually make money on a growing new industry. In other words, you have a chance to clean up while cleaning up.

This book is divided into two parts. The first section, chapter 1, lists forty-five things small business can do to clean up the environment. This is the *USA Today*–style fast read full of quick facts and tips. The rest of the book is an in-depth explanation of the first chapter. We hope your appetite will be whetted by the first chapter and you will follow up by reading the rest of the book.

Questions and Answers with William K. Reilly and Frederic D. Krupp

Environmental Protection Agency administrator William K. Reilly and Environmental Defense Fund executive director Frederic D. Krupp are two people who are very involved with the role of business in environmental cleanup. Reilly advises President Bush and generally represents the White House view on environmental matters. Krupp's organization is a main-stream environmental think tank that challenges government and industry with some of its innovative strategies.

I have included interviews with both of them early in this book to help lay out a more thoughtful "big picture" scenario of what business has been doing and what business should be doing to clean up the environment.

William K. Reilly is administrator of the United States Environmental Protection Agency. He was appointed to his position by President Bush, who has also moved to raise the EPA administrator's job to Cabinet level.

The following are written responses to questions I put to Mr. Reilly in June 1990.

Q. Is it possible in the United States to have both economic growth and solution to pollution? How? Won't profits suffer? Won't people be put out of work?

A. Since 1970, our nation has arrested the most egregious examples of environmental degradation, and made substantial strides in cutting many air and water pollutants. We spent a great deal of money as a nation to do so. Yet it is important to note that during that same period, when significant environmental gains were made, the gross national product increased 70 percent.

I cite this fact for two reasons. First, our environmental commitments have not impeded economic progress; on the contrary, the nation is now growing in a healthier way, and the growth we are experiencing is qualitatively better. Second, it is not fortuitous that great environmental gains occurred during a period of economic activity. Economic growth paid for our environmental commitments; a healthy economy financed our superior environmental performance.

Q. What environmental progress has the United States made since 1970? What failures? What progress will we make in the future?

A. One of the best ways to characterize the progress made is to look at where we were for the first Earth Day in 1970, and where we are on the twentieth anniversary of Earth Day in 1990. On that first Earth Day, I joined other White House staffers to collect trash and debris along the Potomac River. We were very careful not to touch the water. In those days, you were advised to get a tetanus shot if you should happen to fall in the water. Today, on almost any warm day, you can watch wind surfers on the Potomac.

The gains we have made in air and water quality are measurable, significant and indisputable. In most major categories of air pollution, emissions on a national basis have either leveled off or declined since 1970. Emissions of particulates are down 64 percent; sulfur oxides down 25 percent; volatile organics down 29 percent; carbon monoxide down 38 percent; and lead down 96 percent. For the future, our goal increasingly must be pollution prevention. We have a long way to go in many areas.

Q. Isn't a lot of business's newfound interest in the environment just a marketing gimmick?

A. It is no secret that the American public has become more and more aware of the environment, and has also begun to scrutinize the environmental sensitivity of the companies producing products they buy. A good business will recognize this market trend and follow it.

On the other hand, businesses have shown great strides in their investments beyond just pollution control equipment and investments to comply with environmental regulations, which is indicative of an interest in the environment that exceeds the minimum requirements.

Q. What are the benefits to business of a clean environment?

A. When a business wants to attract employees to their facility, a clean environment is a benefit. If the community is heavily polluted, it will ultimately harm the ability of a company to attract and retain employees.

We look, too, at the other side of the equation—the problem of a degraded environment for business. There is a direct connection between prosperity and environmental progress. The countries of Eastern Europe—East Germany, Poland, Hungary, Romania, Czechoslovakia and the Soviet Union—have all suffered the consequences of forgoing pollution control costs. It is interesting that no one can identify any economic "gain" resulting from money "saved" which would have been spent on pollution control.

Q. What can employees, mid-level managers or small business owners do to clean up their own business backyards?

A. In the management decision-making process, the first priority for companies should be to prevent pollution. It makes sense environmentally, economically, and many times it saves the company valuable time and resources. Secondly, the priority should be to recycle or reuse what cannot be prevented economically. Thirdly, treating the waste and finally landfilling or incinerating what is left as a last resort. Companies can encourage their employees as individuals to conserve energy, car pool, plant trees, sponsor cleanups and get involved in educational projects. These volunteer efforts can make a significant contribution to the environment.

Q. Isn't business responsible for much of the pollution we now have? Shouldn't they pay for cleanup out of their own pocket?

A. "Polluters pay" have been very important watchwords for the United States, and I have continued to adhere strongly to that philosophy. And most of the money put into environmental cleanups in the United States is private money: $80 billion or more. Moreover, where we can, like the Superfund, we are trying to recover costs paid by the government. In 1989, for example, we recovered over $157 million and secured about $1 billion in private party cleanups.

Q. Has business been reluctant to take the lead in environmental cleanup? Do you expect this to change?

A. Business is not reluctant. Under Superfund, we have seen an increase in the number of polluter-led site cleanups. Many firms and industries lead the way in preventing pollution and reducing waste. We have seen an increase in the number of polluter-led site cleanups.

Du Pont, 3M, Dow, Monsanto—all have undertaken pollution prevention programs because they believe it will pay off financially in reducing their costs.

Q. Can you point to any examples of good business and a clean environment going hand in hand?

A. 3M has a program called "Pollution Prevention Pays," and Chevron has one called "SMART" (Saves Money and Reduces Toxics). In 1987, the first year of the SMART program, hazardous waste disposal dropped 44 percent from 135,000 to 76,000 tons, saving $3.8 million. Chevron has set a goal of a 65 percent reduction across the board by 1992. Monsanto has a 90 percent goal for reducing toxics.

Clairol, in one plant which produces hair care products, previously flushed their pipes with large quantities of water, wasting the material inside the pipes and generating one thousand tons of liquid hazardous wastes annually. By installing a $50,000 system using a foam ball air-propelled through the pipes to collect the product and clean the system, waste was reduced by 11 percent, saving the company $240,000 a year.

The Borden Chemical Company's Freemont plant near San Francisco reduced the level of toxic organic pollutants in its waste water by 93 percent in the 1980s, using pollution prevention techniques. One major change came when the company changed its method of rinsing large chemical reactor vessels. Borden had been using a single rinse that produced wastewater that was too dilute for recycling but too polluted to dispose of easily. The company switched to a two-stage process: a small first rinse cleans out most of the residual chemicals, which are then recycled; the second rinse produces dilute wastewater, which is readily handled by existing treatment systems. The company is saving money by recycling raw materials that it had been paying to eliminate.

◆

Frederic D. Krupp is the executive director of the Environmental Defense Fund, a not-for-profit environmental group founded in 1967. These are answers to questions posed to him in June 1990.

Q. What role has business been playing in environmental cleanup and what role should they be playing?

A. Many businesses have waited to do environmental cleanup until forced by restrictive laws or court judgments. Fortunately, business is learning it can often offer higher returns to stockholders by taking environmental protection into account from step one. In terms of environmental cleanup, the best role for business is one that prevents pollution in the first place.

Many corporations have acted like "bad guys" on environmental issues in the past. Today, however, more and more are listening to consumers' demands for environmentally sound products and services, and responding. In fact, the stiff competition for environmental dollars is teaching business that in order to stay out of the red, they must operate in the green.

Q. Do you sense a change in the attitude of business about their role in environmental protection?

A. There is a growing environmental ethic that citizens are holding themselves to, as well as their corporate and governmental leaders. Attitudes in business are changing. Public pressure against polluters is one important reason for the changes, but there is a more important change afoot. Business is learning that a healthy environment and a strong economy go hand in hand.

A good example of this is California's Pacific Gas & Electric Company, the largest investor-owned utility in the nation. PG&E had plans to build $20 billion worth of large coal and nuclear power plants. The company thought the plants were indispensable to California's economic health. EDF persuaded PG&E to adopt a package of alternative energy sources and conservation investments instead. How? EDF proved that the alternative package not only met the same electrical needs, but also meant lower prices for consumers and higher returns for PG&E stockholders. With successes like these, it doesn't take long to change attitudes.

Q. In your opinion, can economic growth and a clean environment go hand in hand?

A. Economic growth and a clean environment must go hand in hand. The two are inseparable. The fact is, we will be living in a different world forty to fifty years from now. The question is, How will the world be remade? Right now it is as if the entire world is on a truck careening down the highway toward an unknown future. We can either steer that truck to a clean, safe future, where the industrialized world has freed itself from a hundred-year-long fossil fuel addiction, and where natural ecosystems have survived and thrived. Or we can let it head out of control down the path of continued environmental degradation.

The way that we drive toward the future is not filled with obstacles for business and for the economy, but filled with opportunities. The economic question is not how much it is going to cost to address our problems. The question is which corporations and countries are going to thrive, and which will fail to see the opportunities and falter. If, for example, some car manufacturers don't see that we need very high-mileage cars and whole new clean technologies, then they will not survive into the next century.

Q. Some people say that government bureaucracy and excessive regulation are major stumbling blocks to a cleaner environment. What do you think?

A. Much regulation is necessary. Economic incentives are a supplement—not a replacement—for the environmental standards and controls that continue to be required.

Government bureaucracy and excessive regulation can certainly delay environmental protection. One only need look as far as the nation's toxic chemicals control laws for proof. For twenty years, the country has piled law upon law in its efforts to protect the public from toxics.

Unfortunately, since the first laws were passed, only a handful of chemicals have been adequately regulated. Government agencies assigned to regulate toxic substances often get tied up in seemingly endless wrangles over what level of chemical is unsafe and how it should be controlled. Bureaucrats face political pressures as well as legitimate uncertainty over complex scientific issues. And, until recently, many laws gave and still give companies incentives to drag out the regulatory process.

Why? Because without explicit regulations, it is virtually impossible for the government or individuals to prove that a company's use of a chemical is unsafe. A more effective approach, as demonstrated by California's Proposition 65, rearranges traditional pollution incentives by putting the onus on business to show that its actions are safe, rather than on the overburdened regulators to establish that practices are unsafe.

The effectiveness of this approach is easy to measure. The federal government has effectively regulated only a handful of chemicals in twenty years of effort. In the handful of years Proposition 65 has been in effect, California has successfully set standards for literally dozens of chemicals, and many consumer products have been reformulated in order to meet the law's strict standards.

Q. What progress have we made in environmental cleanup since the first Earth Day twenty years ago?

A. Since the original Earth Day, controls on abusive behavior were created by the passage of the first Clean Air Act, the Clean Water Act and the Toxic Substances Control Act, and the creation of the Environmental Protection Agency. During the Nixon administration, the National Environmental Policy Act, which required that major federal actions be studied for environmental impacts, was signed into law.

Legislative gains actually continued in the 1980s, after Love Canal helped turn the federal focus to the safe disposal and handling of toxic substances. During the Reagan administration, EPA administrator Anne Gorsuch Burford and secretary of the interior James Watt largely failed to roll back existing environmental protections and their efforts to do so made environmental groups' membership grow. This membership growth led to the general public awareness of environmental problems that helped set the stage for Earth Day 1990.

1

Forty-five Steps Business Can Take to Clean Up the Environment

Smaller businesses have particular problems when they set out to do their part in cleaning up the environment. In most cases, they have to comply with federal, state and local environmental regulations even though they don't have the extensive legal departments or chemical engineers on staff to sort through the cascade of restrictions.

Furthermore, owners and employees are about as busy as they can possibly be trying to run their shop, leaving little time to research ways to cut down on the amount of pollution, even though it may be small, they have a hand in creating.

Businesses, especially small businesses, are made up of people. People who care just as deeply about environmental affairs as individuals, groups and agencies.

This chapter is designed to give small business tips, techniques and strategies for cleaning up your business's backyard. I have put in many examples of small business activity in order to make all this information relevant to your situation.

I hope I have helped. Good luck!

Administrative

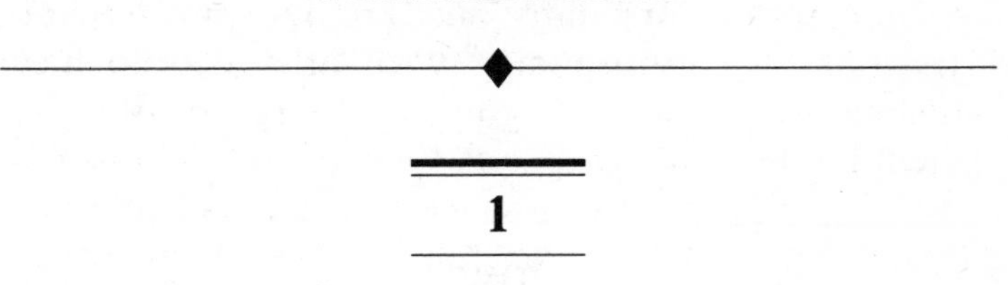

1

Appoint an environmental coordinator.

WHY?

- Business in the 1990s should be prepared to respond to environmental concerns of customers, stockholders, consumer groups and employees.
- Regulatory agencies and state and local governments are increasingly putting new restrictions on business activity as it relates to the environment. Example: Suffolk County, New York, and Hennepin County, Minnesota, have placed bans or restrictions on certain types of plastics. If you are a business using plastics in packaging or elsewhere, you need to know what your alternatives are and how you can respond to these pressures.
- You need a person to research and coordinate new office recycling programs, to make purchasing decisions for recycled products.
- There is a lot of confusing and conflicting environmental information available today. The environmental coordinator can sift through this and present good suggestions.
- Liabilities and fines are starting to be levied on business for being out of compliance with laws and regulations. It is better to be ahead of this curve rather than behind it.
- To avoid potential problems and costly mistakes. Example: You are planning to buy a new building

to house your expanding small business. Your environmental coordinator needs to inform you that if that site is contaminated with fifty-year-old hazardous waste, you are going to be responsible for cleaning it up.

HOW!

- The person you designate as environmental coordinator must have or be given authority to speak internally for the company on environmental matters. You should set company policy, but you should support your enviro-coordinator once that policy is set.
- Choose a person who knows how to do research. You don't have to be a chemical engineer to be an enviro-coordinator, just must be able to read, listen and learn the proper steps.

2

Develop a companywide environmental policy.

WHY?

- A companywide policy sets standards that everyone can agree upon.
- When an environmental question arises and there is no clear-cut answer, you can refer to the policy for guidance.

HOW!

- Get the facts. There is no reason to reinvent the wheel. Contact your state or local recycling agency, local or national environmental organization or business association for information already on file.
- Get your employees involved. Tell each one you are developing a policy and ask if they have any tips or ideas that should be considered.
- Read the *Valdez* Principles, developed by the Coalition for Environmentally Responsible Economies, on page 161–64 as an example of company policy that can be adapted or modified to suit your needs.

3

Recycle your office paper.

WHY?

- Each office worker in the United States produces an average of 1.5 pounds of office trash each day, according to the San Francisco Recycling Program. As much as 90 percent of business trash is recyclable office paper, meaning each office worker in the United States throws away as much as 337 pounds of paper each year that could be recycled.
- For each ton of office paper we recycle, seventeen trees are saved from the ax, and three cubic yards of landfill space are not filled up.
- Recycled office paper can be made into other high-quality office paper, as well as napkins and hand towels.
- Recycled office paper is a valuable commodity. Recyclers are paying over $200 a ton for premium grades of paper. If your office alone or in concert with other offices in your building can generate enough tons of office paper, you can make money, or at least lower your carting costs.
- Office paper recycling is becoming mandatory in cities such as Washington, D.C., and New York. It will soon be coming to your town. You might as well get started now.

HOW!

- Appoint a person who will be responsible for research and development of your program.
- Call your city or state recycling coordinator to get advice.
- Research recyclers listed in the phone book to find one who will handle the amount of office paper you generate.
- Buy and install a dual wastebasket system by each employee's workstation—one for recyclable office paper and one for all other trash. Your normal paper supplier, office equipment supplier or even your trash carting company will probably be able to supply you with equipment. But, really, any small-sized wastebasket or bin for around the desk, plus larger-sized receptacles for collection, will work. Diversified Recycling Systems, 5606 North Highway 169, New Hope, MN 55428, is a manufacturer and distributor of office recycling products. For a dealer near you, call 612-536-6662. For a very small office, contact Seventh Generation, Colchester, VT 05446-1672, 800-456-1177, for their catalogue that sells office paper recycling bins.

According to Tom Lewis, general manager of the Johnston Paper Company in Auburn, New York, it will cost approximately $200 to equip an office of ten people with office paper recycling containers. The problem is that small offices do not generate enough paper to make it profitable for large recyclers to make pickups at their site. The solution is mandated citywide office paper recycling by existing sanitation departments, which will take whatever you leave them. In the meantime, you are going to have to haul it yourself to a local municipal transfer station or work in concert with other offices in your building to generate enough paper to entice a hauler.

Johnston Paper Company has started selling roll-away tilt trucks that hold two thousand pounds of paper to small businesses. When these trucks are full, Johnston collects the paper, pays the small business a price, bales the paper back at their own site and sells it to larger haulers. It is a way to encourage sales of their own recycling supplies but also a way to clean up the environment. Johnston launched this program in April 1990. Their own in-house office paper recycling program collects two thousand pounds every two weeks. They have thirty-five employees.

- Read chapter 2 of this book for profiles of office paper recycling systems at AT&T in New Jersey, Waste Management, Inc., in suburban Chicago, Fort Howard Paper Company in Green Bay, Wisconsin.

4

Set up an office recycling center for newspapers, magazines, cans and bottles.

WHY?

- Even the smallest office can collect dozens of soft drink bottles and cans, plastic and paper take-out food packaging, newspapers, magazines and other recyclables every week.

HOW!

- Place large bins for recyclables in a heavily trafficked area in your workplace. Near the canteen or employee lounge or next to soft drink or candy dispensers are good places.
- Learn what your sanitation department or recycling centers will recycle and how carefully things need to be separated.

5

Conserve paper use.

WHY?

- Cutting down on the amount of paper you use can save trees and reduce the amount of paper you have to recycle or send to the landfill.

HOW!

- Print material on both sides of the sheet of paper when making copies. You can cut your copying costs for paper by 50 percent.
- Use scratch paper for in-house memos.
- Post general all-company memos in a few well-trafficked locations rather than making individual memos for each employee.
- Post memos electronically on your integrated office computer bulletin board system.

Supplies

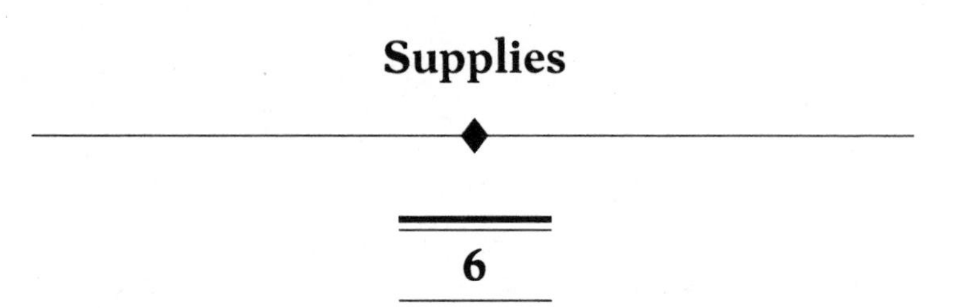

6

Buy stationery, envelopes and other office paper made from recycled paper, preferably postconsumer office paper.

WHY?

- To encourage the development of more office paper recycling programs and reduce the number of trees that have to be cut down each year.

  The reason? Most recycled paper is made from trimmings collected in the printing industry. This is indeed recycled paper, and it is better to use this paper than order stock made from virgin wood. Postconsumer recycled paper is stock made from used paper collected from office recycling programs.
- The federal government requires that businesses that receive federal money and use more than $10,000 of federal money to purchase paper and office supplies must stipulate a preference for postconsumer paper.

  CAVEAT: Postconsumer paper is 5 to 15 percent more expensive than ordinary recycled paper, which is roughly the same price as virgin paper.

HOW!

- Ask your existing office paper supplier to demonstrate their product line of recycled products.

- For the very small office, the Seventh Generation catalogue (800-456-1177) has a line of recycled office paper, including four-by-four-inch desk pads, eight-and-a-half-by-eleven lined pads, computer paper, copier paper and telephone message pads.
- Or you might contact these companies:
  - Quill Corporation, 100 S. Schelter Road, P.O. Box 4700, Lincolnshire, IL 60197-4700, 708-634-4800, for east of the Rockies.
  - Quill Corporation, 5440 E. Francis Street, P.O. Box 50-050, Ontario, CA 91761-1050, 714-988-3200.
  - Earth Care Paper Company, P.O. Box 3335, Madison, WI 53704, 608-256-5522. Specializes in small orders.
  - Conservatree, 10 Lombard Street, San Francisco, CA 94111, 415-433-1000. Outside California, 800-522-9200.

7

Buy ink pens, tape dispensers and other office products that are reusable, refillable or made of recycled fibers.

WHY?

- American offices throw away tens of thousands of office pens and products that simply end up in landfills.
- Buying items made of recycled fibers helps encourage the growth of the recycling market.

HOW!

- Ask your suppliers to sell you ink pens and refills. They cost more up front but last longer and cut down on trash.
- Contact the Pimby Company, P.O. Box 240, Purdys, NY 10578, 914-277-8872, for pens, rulers, scissors and other office tools made from recycled plastic.

8

Use nontoxic fluids and art supplies in your graphic arts department and for general office use.

WHY?

- Permanent felt-tip markers, rubber cement, epoxy, spray fixatives, dye, inks and other common art supplies contain hazardous solvents and powders that when inhaled or passed through the skin cause health problems.
- The Center for Safety in the Arts says that artists are beginning to suffer from cancers, miscarriages, nervous system damage and other ailments as a result of using toxic and irritating art supplies.

HOW!

- Buy the For Everything brand correction fluid by Wite-Out, Inc., because it does not contain trichloroethylene or the ozone-depleting chemical methyl chloroform.
- Be sure the art department is adequately ventilated to remove any damaging vapors.

- Use water-based glues, markers, inks and paints instead of those based on solvents.
- Contact the Center for Safety in the Arts, 5 Beekman Street, New York, NY 10038, 212-227-6220. They offer workshops, videos and other educational material on risks and alternatives to hazardous art supplies.

The Office Kitchen, Lounge and Bathrooms

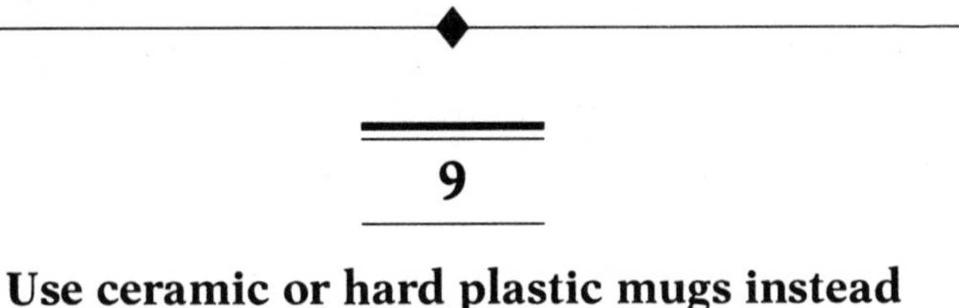

9

Use ceramic or hard plastic mugs instead of Styrofoam or paper cups for coffee and other beverages.

WHY?

- Paper cups cannot be recycled. Polystyrene cups can be recycled, but the availability of polystyrene recyling programs is very limited. The end result is that both paper and Styrofoam cups end up in the trash and ultimately in the landfill.

HOW!

- Give or sell to each employee a reusable ceramic mug, either plain or adorned with the company logo or with a favorite recycling slogan.
- All the mugs are rounded up each night by individual employees or the janitorial staff and washed in the dishwasher.
- Ask employees to bring their own private mug from home that they are responsible for washing and keeping clean.

FACT: Home Box Office (HBO) in New York, with eleven hundred employees, cut their monthly coffee cost by $450 when they switched to HBO reusable mugs. It seems that people had a habit of walking by the coffee pot, grabbing a Styrofoam cup, pouring a cup of coffee, returning to their desk and letting it go cold. They would then run out and repeat the process. With the new mugs, people became more conscious of their coffee habits and more conservative in pouring cups.

10

Use unbleached coffee filters.

WHY?

- The poison dioxin has been found to be present in paper coffee filters bleached white at the paper mill.

HOW!

- Switch to unbleached coffee filters sold by Melitta and other companies. The price is the same as bleached.

11

Install a modern energy-efficient dishwasher and refrigerator.

WHY?

- New federally mandated energy efficiency standards, which take effect in 1990, will make major appliances 10 to 30 percent more efficient than earlier models.

- More energy-efficient appliances means a lessening of demand for electricity, meaning fewer costly new power plants and less air pollution.

HOW!

- Contact your local electric utility for information on how to buy the most energy efficient appliances.
- Carefully read the EnergyGuide label on the front of the appliance to determine which unit suits your needs and is the most energy efficient.

12

Pack your lunch in reusable hard plastic containers or aluminum foil.

WHY?

- Waxed paper and plastic wrap cannot be recycled and they are almost impossible to reuse. They end up at the landfill.

HOW!

- Buy hard plastic Tupperware or Rubbermaid containers. Put sandwiches, salads, vegetables, etc., in them to take to work. They can be kept cold in the refrigerator and heated in the microwave with no additional packaging.
- Instead of packing throwaway juice or beverage containers, pour liquid in a small hard plastic or Thermos-type bottle or jar.
- Toss everything in the dishwasher at home at night.

- Pack lunch in a colorful and reusable lunch pail instead of a fresh paper bag. If you do prefer paper lunch sacks, at least reuse it several times until you recycle it.

13

Buy hand towels, napkins and toilet paper made from recycled paper.

WHY?

- These products are readily available and increasingly being made from recycled office paper. They offer the same comfort and quality as products made from virgin paper.

HOW!

- Environ-brand hand towels, napkins and toilet tissue are made and offered to commercial customers by Fort Howard Paper Company of Green Bay. For smaller businesses, tell your supplier you want products made from recycled paper or simply buy these items in the grocery store. Marcal products, and many others, are made from recycled paper and widely available in grocery stores everywhere.

14

Install water-conserving aerators on faucets and place a dam in toilet tanks.

WHY?

- Aerators can reduce water usage by 50 to 75 percent without any loss of washing potential.
- A toilet flush typically uses five to seven gallons of water. You can reduce that amount by 25 to 40 percent with a toilet dam.
- Fresh water is a scarce resource and it is expensive to purify and transport through pipes.

HOW!

- Most modern kitchen faucets come with aerators, but they can be purchased at hardware stores for under $5.
- A toilet dam can be made by filling a one-quart plastic soft drink bottle with water, screwing the cap on tightly and placing it in the toilet tank. Be sure not to interfere with the flushing mechanism.
- Don't use bricks in your toilet tanks because they tend to be dirty and can clog up the water flow.

15

Avoid air fresheners in toilets.

WHY?

- Most air fresheners do not really freshen the air. They emit a chemical vapor that masks other odors.

HOW?

- Prevent odors in the first place by keeping toilets and food areas clean and well ventilated.
- Use alternative air fresheners such as potpourri or flowers.

16

Use nontoxic janitorial cleaners and supplies.

WHY?

- Many commonly used disinfectants are actually classified as pesticides by the EPA. They are considered a hazardous waste.
- Many commonly used cleaners contain ammonia, chlorine or harsh detergents, and may contain solvents. All of these ingredients are considered low-level hazardous waste because they are either caustic or toxic.

HOW!

- Switch to alternatives such as Murphy's Oil Soap, Spic and Span or plain soap and water.
- Use baking soda or soap and water as scouring agents for tiles and sinks.

Clean Air at Work

17

Have an air quality audit done on your office.

WHY?

- The EPA now says that indoor air pollution is more hazardous to your health than outdoor air pollution.
- Nearly one-third of all new commercial buildings have polluted air.
- As many as 20 percent of office workers are suffering from "sick office syndrome," a health condition that is caused by inadequate ventilation combined with fumes, particles and other contaminants commonly found in office building air.
- The symptoms of sick office syndrome could be headache, dizziness, fatigue, sore throats and eyes, colds and flulike problems.
- Sick office syndrome will cost business $10 billion a year in lower productivity, sick days and direct medical costs, according to the EPA.
- Legionnaires' disease, which is a bacteria spread through air conditioning systems in large buildings, killed twenty-nine people at a convention in Philadelphia in 1976, and recently killed ten people over a period of three years at a hospital in South Dakota.
- The *Wall Street Journal* reported that over a dozen

indoor air pollution suits have been filed in the last two years and more are expected.

- Sick office syndrome is believed to have emerged in the late 1970s, when buildings were heavily weatherized and sealed to save fuel during the energy crisis.

HOW!

- Have your heating, cooling and ventilation system inspected by a licensed contractor or by your building management. Most sick office syndrome problems are cured by replacing air filters, repairing faulty systems and increasing ventilation flow.
- The number of indoor air quality experts is growing rapidly and none of them is licensed or regulated by government. Most of these experts are honest and well informed, but buyer beware. Ask for and check references carefully.
- Be sure that they are checking for formaldehyde and other gases, bacteria and fungi.

18

Implement a no-smoking policy in the workplace.

WHY?

- Passive smoking causes 3,800 cancer deaths a year, according to the EPA.
- Passive smoking causes 50,000 deaths a year, two-thirds of them from heart disease, according to Dr. Stanton Glantz of the University of California at San Francisco.

HOW!

- Inform and alert all employees that you will be issuing a no-smoking policy in the workplace and announce when it will take effect. Three to six months' warning is time enough for them to be prepared.
- Encourage smokers to get help from their physicians or local lung association.

19

Have a lead and asbestos hazard check done on your office if you are in an old building.

WHY?

- Lead in paint for interior use was not banned nationwide until 1977. The U.S. Health Service estimates that 3 million tons of lead are still in old paint around the country. Lead in paint becomes a health threat during removal or remodeling, when sanding or scraping releases lead dust and particles.
- In children under six, lead poisoning can cause learning disabilities and even permanent brain damage. Parents working in an old office undergoing remodeling can bring lead dust home on their clothes and contaminate their young children.
- Asbestos released in older deteriorating buildings can get spread through a ventilation system and cause irritations and even illnesses.

HOW!

- Asbestos and lead paint in good condition that is not deteriorating should be left alone. Removal of both lead and asbestos is expensive, very dangerous and should only be handled by professionals.
- Contact your department of health or environmental protection to conduct an inspection or ask for references from them of people they approve as inspectors.

20

Keep office air fresh and well ventilated.

WHY?

- The main cause of sick office syndrome is stagnant air (i.e., lack of fresh air circulating through an office that will dilute and remove any airborne contaminants).
- Oftentimes air will become stagnant as a result of office reconfiguration (i.e., building room dividers and cubicles without compensating for airflow into these areas).
- It is not uncommon for air intake vents to be clogged or partially covered, or located near a garage or street that allows intake of carbon monoxide from exhaust.

HOW!

- Have your ventilation sytem inspected by a licensed contractor or building superintendent.
- Be sure fresh air intake vents are clean and free of obstructions.

- Install direct ventilation for graphics arts department or other area where solvents and chemicals are used.

21

Remodel and renovate your office with healthy furniture and decor.

WHY?

- Many offices are furnished with synthetic woods, carpeting and room dividers that give off formaldehyde vapors and are held together with glues and adhesives and painted with paints that give off volatile organic gases that contribute to sick office syndrome.

HOW!

- Switch to natural fiber carpets, upholstery and drapery.
- Install metal or hardwood desks and cabinets instead of those made from particle board.
- Use water-based latex rather than oil-based paints and solvents when painting.
- Natural wood and fiber furnishings do cost more, as much as 25 percent.
- If for economic reasons, you still need to furnish your office with synthetic items, be sure to fully ventilate the office, preferably over an entire weekend with maximum airflow to remove vapors before you allow workers to return.

Maintenance—Energy and Lighting Efficiency

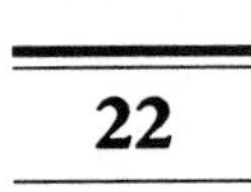

Reduce air pollution through energy-efficient heating and cooling.

WHY?

- Acid rain and the greenhouse effect leading to global warming are results of burning fossil fuels, especially those to heat and cool our buildings.
- Exploration, production and transportation of fuel oil, coal and natural gas cause environmental damage in the form of spills, leaks, abandoned strip mines and other problems.
- Oil, coal and gas are vast but still limited natural resources that should be used wisely and frugally.

HOW?

- Conduct an energy audit on your office or building. Most utilities will either conduct the audit for you at a modest cost or refer you to energy auditors who specialize in commercial property.
- Seal air cracks and leaks around windows and doors at the workplace, just as you would at home. Insulate where necessary.
- Have your heating and cooling systems inspected and tuned up each year if you own your building. This will cost from $50 to $100 and include cleaning, installing new filters, checking for leaks and

adjustment. A tune-up will pay for itself every year in saved energy costs and prolonged life of equipment. An inefficient system can waste as much as 50 percent of energy costs.

- Turn down the thermostat to 65 to 70 degrees in winter and turn up the thermostat to 78 degrees in summer. For every degree you set your thermostat back, you can save 1 to 3 percent of your heating and cooling costs. Be sure to set the thermostats back when you leave the building at night.

  For example, 3M invested $11,800 to install setback thermostats in office areas to maintain a nighttime temperature of 55°F. The reduced demand in heating and cooling saves 3M $56,000 annually in energy costs.

- Install energy-efficient heating and air conditioning systems. Contact your local utility for cost comparisons and savings estimates comparing your existing heating and cooling systems with more energy-efficient modern ones. Many utilities are so sold on the economic and environmental value of high-efficiency heating and cooling systems, they are offering incentives and rebates for commercial customers.

  For example, Pacific Gas & Electric, in California, is telling commercial customers they can achieve energy savings of up to 50 percent with more efficient heating and cooling systems, and that investment can be paid back in as little as two years. In 1990, they were offering incentives of up to 50 percent of the cost of new equipment to commercial customers.

23

Maximize efficiency of workplace lighting, prevent pollution and save money.

WHY?

- Energy-efficient lighting can save business 20 to 50 percent a year on lighting costs.
- Lighting is powered by electricity, which is generated primarily in the United States by burning coal, oil and natural gas. The burning of coal and fuel oil emits sulfur dioxide, a prime component of acid rain, as well as carbon dioxide, a most damaging greenhouse gas.
- One high-efficiency light bulb can eliminate the combustion of 220 to 382 pounds of carbon, a source of greenhouse gases, according to the President's Council on Environmental Quality.

HOW!

- Conduct a lighting audit on your workplace. Check to see if areas are over- or underlighted, if your bulbs are older, more inefficient varieties, if fixtures are clean, if you are lighting areas that don't need it.
- Turn off lights when you leave the workplace at night and when you go out to lunch.
- Clean fixtures and remove dust and dirt, which can cut efficiency by 30 percent.
- Remove unnecessary light bulbs and disconnect ballasts in fluorescent fixtures, especially in lights

near windows or in seldom-used hallways and corridors. Install lower-wattage lamps in areas that are overlighted. Employees can suffer eye strain and fatigue from glare caused by too much light. Use your judgment and try reducing overhead light.

- Install high-efficiency, long-lasting fluorescent and incandescent lamps when replacing burned-out ones. For example, installing a 67-watt high-efficiency incandescent bulb for an existing 75-watt bulb uses 11 percent less electricity and yields only 5 percent less light.
- Install compact fluorescent light bulbs in the place of incandescent. They cost about $15 each but last thirteen times as long as a regular incandescent.
- Take advantage of natural lighting. Many offices that are located near windows can reduce their lighting by half. Public areas, reception rooms and corridors located near windows or skylights may need no artificial lighting at all.

24

Landscape your grounds and building for energy efficiency.

WHY?

- Shade provided by trees can lower ambient air temperature by 10 percent. Lowered air temperature outside your building can mean less demand for cooling inside—from 10 to 50 percent—lowering energy costs.
- The Missouri Division of Energy in Jefferson City feels so strongly about the value of landscaping to

hold down energy use they awarded $500,000 in grants for energy-efficient landscaping projects around schools and government buildings in 1990.

- An average tree removes thirteen pounds of the greenhouse gas carbon dioxide from the atmosphere each year.

HOW!

- Contact your local nursery or professional landscapers for expert advice on tree and bush plantings for energy efficiency.
- Large trees can cost as much as $10 a foot plus planting and maintenance fees.
- Be sure to plant trees that lose their leaves in winter so that warm solar rays can penetrate your windows.
- If you can't plant trees, try installing canvas, metal or hard plastic awnings on windows to block heat during the cooling season.

Transportation

Commuting back and forth to work creates 30 to 35 percent of the polluted air and smog we have to breathe. The stress caused by traffic congestion can lower worker productivity and make it more difficult for business to recruit and keep qualified workers.

25

Walk or bike to work.

WHY?

- Over one-half of the people in America live within six miles of where they work. Where it is practical, walking or riding a bike to work is a good alternative to congested streets.

HOW!

- Install secure bike racks or storage areas at the work site to encourage bicycling.
- Support the creation of biking and walking trails built out of old railroad beds and other paths. Seattle now has thirty miles of trails that are used by as many as ten thousand commuters a day to get back and forth to work.

26

Take public transportation and subsidize employees who do.

- Every commuter who takes public transportation prevents the release of sixty-three pounds of carbon monoxide, nine pounds of smog-producing hydrocarbons and nitrogen oxides and one pound of soot into the atmosphere.

HOW!

- Offer to pay employees $15 a month if they will take public transit to work. In Los Angeles, companies with more than one hundred employees that offer subsidized parking are required to do this. Smaller companies can receive $5 a month from the city of Los Angeles to subsidize employees who commute by public transit. You could use this offer as a benefits package incentive to attract and keep good employees.

27

Develop incentives for car and van pooling.

WHY?

- The average commuter car is now carrying only 1.3 people.
- Air pollution is now worsening because the number of cars on the road is increasing.

- It is extremely expensive and difficult to build more roads, especially in congested urban areas.

HOW!

- Establish preferred parking areas for employees who car pool.
- Contact your state or local transportation department and get the car pool information they have. Make information available to your employees.
- Give employees time at work to arrange car pools with other employees.

28

Maintain and inspect all company vehicles for fuel efficiency and emissions reduction.

WHY?

- A property tuned automobile can save 5 to 15 percent in gas usage each year and reduce carbon monoxide and hydrocarbon emissions by 25 to 30 percent, according to the American Lung Association.
- Improperly discarded motor oil, batteries, tires and other fluids cause water and soil pollution and can contribute to fires and spread of disease.
- A poorly maintained and serviced air conditioner emits ozone-depleting CFCs.

HOW!

- Conduct an annual inspection of all company vehicles, checking for fuel efficiency and smog-producing hydrocarbon emissions.
- Ask to make sure that your service station or department is recycling spent motor oil, used batteries and tires.
- Use your air conditioner frugally and ask your service station or department to recycle the Freon from the air conditioner during servicing.

29

Buy fuel-efficient company cars.

WHY?

- The fewer miles per gallon your car gets, the more greenhouse gases it gives off. For example, if your car gets 18 mpg, it gives off 57.7 tons of carbon dioxide over its lifetime. If it gets 27.5 mpg, it gives off only 37.71 tons of CO_2 over its lifetime.
- Higher fuel efficiency means lower gasoline expenses.

HOW!

- Get the free *Gas Mileage Guide,* Consumer Information Center, Pueblo, CO 81009.
- Specify that you want all company cars to exceed a certain fuel efficiency standard. The state of New York just raised their fuel efficiency requirements for state-owned automobiles to 29.5 mpg from 27.5.
- When you go to buy a car, tell the dealer you want

to know how many miles per gallon it gets in town and on the highway.

- Choose the car with the highest miles per gallon that still suits your company needs.

30

Fuel your cars with clean gas, reformulated or ethanol.

WHY?

- Reformulated gas contains an oxygen-rich additive that makes the fuel burn more cleanly, giving off less carbon monoxide and less smog-producing hydrocarbons.
- Ethanol contains a grain alcohol additive that gives off less carbon monoxide and less carbon dioxide.

HOW!

- If you have a pre–catalytic converter car, buy ARCO EC-1 in Southern California. For leaded gas cars, buy Shell SU 2000E or Marathon's Amaraclean gasoline. Prices for gasoline are regional and unpredictable, but you can expect to pay either the same price or slightly more for reformulated gas.
- Ethanol is more common in the midwestern states. Look for an ear of corn emblem on the pumps and ask for gasohol.

31

Buy alternate fuel vehicles.

WHY?

- Natural gas is a very clean-burning fuel that gives off very little carbon monoxide or smog-producing hydrocarbons.
- Electric cars give off no direct pollution and utilities say they would not have to build new power plants to supply the needed electricity for automobiles.

HOW!

- Contact your local natural gas company and tell them you are interested. United Parcel Service is now converting its 2,700 vans to natural gas in Los Angeles.
- General Motors will be test marketing 1,000 trucks powered by natural gas in early 1991.
- General Motors, Ford and Chrysler are all working to develop electric cars.

32

Enforce a policy of no idling delivery cars and trucks.

WHY?

- Leaving your car idling for more than thirty seconds, instead of turning it off and restarting it, does not save gas. It is simply spewing hydrocarbons and other contaminants into the atmosphere and wasting gas.

HOW!

- First, inform employees of the new policy and explain to them you are doing it to clean up the environment and save gas.
- New York City has now sent special traffic police out to monitor idling vehicles and they are handing out summonses to offenders. Inform your employees that you will also fine any employee who gets a ticket for idling, and that the money will go into an environmental activity like tree planting or recycling.

Packing and Packaging

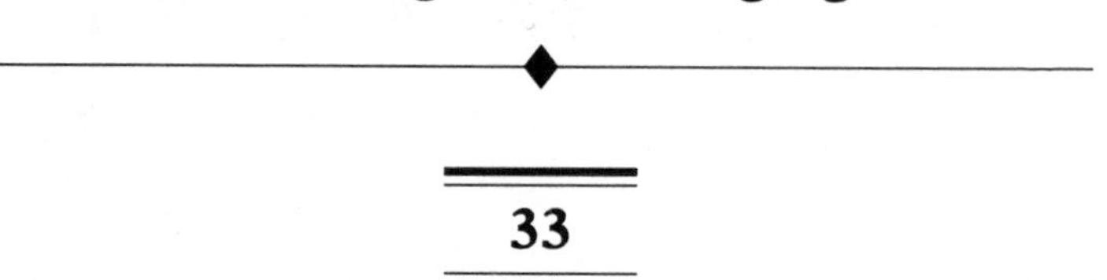

33

Change packing materials to cut down on waste.

WHY?

- In most cases, packing material, both plastic and paper, is simply discarded as waste by your customers, who are increasingly aware and angry about polluting packing. It will end up in landfills and contribute to our solid waste problem.

HOW!

- Use packing paper products made from recycled paper fibers. This includes jiffy packs, all types of envelopes, tissue paper, cartons and boxes. Ask your distributor.
- Recycle plastic packing such as polystyrene nuggets and bubble wrap that you receive and is in good condition.
- Stuff cartons with recycled newspapers as filler.
- Berry-Hill, a small mail order firm in Ontario, started using popcorn as packing instead of polystyrene nuggets.
- Include a note in your packages telling your customers you are using packing made of recycled fibers. Include information telling them how they can recycle the packing you just sent them.

34

Eliminate unnecessary packaging around your products.

WHY?

- Thirty to thirty-five percent of all trash that goes to the landfill is packaging.
- Seventy-eight percent of Americans are willing to pay more for a product packaged with recycled or biodegradable materials, according to a poll taken for the Michael Peters Group, a New York packaging consulting firm.

HOW!

- Streamline your packaging. McDonald's Corporation has cut over 1 million pounds of waste by shortening its straws and reducing the weight of some of its packaging.
- Make your packaging out of recycled materials. Procter & Gamble makes its plastic detergent jugs out of 25% recycled plastic. Anheuser Busch stipulates that its beer bottles must be made of 30 percent recycled glass. Kellogg's and other cereal manufacturers make their cartons out of recycled paper. Ask your distributors.
- Make your packaging out of materials that can be recycled in your sales area. Technically, anything can be recycled, but few areas are set up to recycle plastic and combinations of plastic, paper and metal.

Manufacturing

Make your manufacturing process as pollution-free and energy efficient as possible.

WHY?

- Several states, including California and Massachusetts, have passed laws regulating and banning certain solvents, petrochemicals, heavy metals, chlorofluorocarbons and toxins in products now being manufactured.
- Rather than paying heavy fines, small business should look for ways to eliminate polluting chemicals from their manufacturing processes.
- Handling and disposing of hazardous waste and polluting emissions is a much more difficult and expensive process than preventing pollution in the first place.

HOW!

35

Conduct an environmental audit of your product line.

- Eliminate mercury and other heavy metals from the inks and pigments you use in printing and packaging.
- Use water-based solvents whenever possible, eliminating chemical solvents.

- Look for ways to use recycled glass, plastics, metals and paper products in your products.

  For example:
  - Gloucester Company, Inc., of Franklin, Massachusetts, was being fined $1,400 a year for manufacturing its caulk with chemicals targeted by the state's toxic chemical reduction plan. They eliminated the offending chemicals and were given an award for environmental achievement.
  - American White Cross Laboratories, Inc., of New Rochelle, New York, found a way to alter its manufacturing process to cut by 50 percent the amount of paper used to make their yearly 1 billion adhesive bandages they sell to drugstores.

36

Install more efficient motors and fans in manufacturing processes.

HOW!

- Contact your local electric utility or fan and motor distributor. Show them your annual energy costs spent on running your existing fans and motors. Let them cost out the benefits of buying and installing new and more efficient fans and motors.
- In 1981, 3M replaced a standard 100-horsepower motor on a furnace with an efficient 20 hp motor. Cost: $817. Savings: $338 a year.

37

Look for ways to modify your production processes to eliminate toxic or hazardous waste pollution before it is created.

HOW!

- Move underground storage tanks of fuels and chemicals aboveground to eliminate leakage. Underground storage tanks, primarily full of oil, gasoline and industrial chemicals, can corrode and crack, spilling toxic fluid into the ground, which eventually seeps into the water supply.

  For example, Wood-Kote Products, Inc., of Portland, Oregon, invested $250,000 to build aboveground storage tanks needed to hold materials for their wood stain and finishes products.
- Recycle and reuse heat and solvents.

  For example, American White Cross Laboratories invested $350,000 in a thermal oxidizer that recycles solvent vapors from their adhesives back into the production process as fuel for their ovens. They have eliminated their sixty thousand gallons of annual hazardous waste that they had to pay to dispose of.

Environmental Investing

38

Consider investing company pension and employee savings plans and/or company profits in environmental stocks and mutual funds.

WHY?

- Business, government and institutions will be spending billions of dollars in the 1990s to comply with new federal, state and local environmental regulations and to clean up environmental damage that has already occurred.
- Companies that are specializing in solid and hazardous waste management, pollution monitoring and abatement, recycling, creating new products out of recycled materials, environmental auditing and engineering, and others, are considered by many investment experts as having great potential for growth in the 1990s and beyond.
- Your investment dollars, combined with other environmentally aware investors', can have a positive impact on a company's environmental policies and performance.

HOW!

- Contact your established broker or these specialists for prospectus information. There are many of these funds and this list is only a sampling and not a recommendation. Many of the major brokerages have funds and advisers available to you. For example:

- Freedom Environmental Fund, One Beacon Street, Boston, MA 02108, 617-523-3170 or 800-225-6258.
- Fidelity Select Environmental Services, The Fidelity Building, 82 Devonshire Street, Boston, MA 02109.
- Environmental Awareness Fund, 1016 W. 8th Avenue, King of Prussia, PA 19406, 800-523-2044.
- Working Assets, 230 California Street, San Francisco, CA 94111, 800-533-3863.
- Calvert Social Investment Fund, 1700 Pennsylvania Avenue, NW, Washington, D.C. 20006, 800-638-6731.

39

Use company credit cards, checking accounts and telephone systems that donate a portion of their proceeds to support environmental causes.

WHY?

- This is a painless way at no cost to you to make a donation supporting your concern for the environment. Typically .5 to 1 percent of the money that flows through your phone or credit card account, or a small percentage of your checking account fees, will be donated to an environmental group.
- You can make a public statement about your concern for the environment because credit cards and

checks are brightly colored and generally printed with pictures of animals and/or the earth.

40

Research and develop ways your company can start new business activity based on environmental cleanup.

WHY?

- One hundred billion dollars was spent on environmental cleanup in 1989, according to the National Association of Manufacturers. Hundreds of billions more will be spent in the next decade.

HOW!

- Examine what your company makes or what services it offers. What is your specialty and how can you put it to use cleaning up the environment?
- There will be tremendous demand for consumer goods that do not create hazardous waste or excess trash, for consulting services in environmental chemistry and engineering, for companies that can transform the millions of tons of garbage we produce into usable goods, for legal and other advice on compliance with government regulations, for energy and transportation management systems, for technology relating to monitoring and cleaning up our air, water and soil.
- The Church & Dwight Company, makers of Arm & Hammer baking soda, found themselves being touted as producing an environmentally friendly cleaning product. They have started marketing

Arm & Hammer deodorants, skin-care items, mouthwash, a paint-stripping compound and a system for removing lead from water systems.

- EmTech Environmental Services of Fort Worth, Texas, found that by spraying oil-eating microbes on oil spills they could contain and clean up tanker spills.

Travel and Entertainment

41

Cut the amount of pollution you create while traveling and eating out on business.

WHY?

- Thousands of people travel on business every day, creating tons of garbage that is often single service or convenience-oriented and that is difficult to recycle.
- Thousands of gallons of fuel are burned each year for business travel. This could be reduced.
- Tons of paper are used at business meetings and conferences each year. This could be recycled.

HOW!

- Stay in hotels that have recycling programs. The Hyatt Regency in Chicago is now processing 140,000 pounds of waste each month. Their program was set up by ReCycleCo., Inc., in Chicago.
- Fly on airlines that have recycling programs.
- Bring your toiletries along with you, packed in reusable hard plastic containers for toothbrush, shampoo, soap, talc, etc.
- When traveling on business by car, carry two plastic or paper litter bags—one for recyclables like beverage containers, the other for trash. An increas-

ing number of gas stations have separate bins for recyclables and other trash.

- Reuse or carry your own conference name tag.
- Encourage the conference center to have a recycling program for the reams of high-quality paper that are not used at the conference.
- Have your suits and dresses pressed occasionally instead of dry cleaned, because dry cleaning uses solvents that when released into the atmosphere contribute to ozone, a key component of smog.

42

Cut the trash while eating out.

WHY?

- Thirty-five to 40 percent of the American food dollar is spent in restaurants, more than half of that by business people.
- Restaurants contribute enormous amounts of glass bottles and cans, food scraps, cardboard boxes, paper and plastic wrappings and other waste that end up in our landfills.

HOW!

- Eat in restaurants that have recycling programs. Hard Rock Cafes in many locations have started recycling glass, cardboard, paper and aluminum. They use only degradable and recyclable serving containers and they donate money to environmental causes. Four hundred fifty McDonald's restaurants in New England have started recycling their Styrofoam packages.

Restaurateurs can contact the National Restaurant Association, 1200 17th Street, NW, Washington, D.C. 20036-3097, 202-331-5900, for more information.

- Don't overorder and waste food. Food waste from restaurants, institutions and homes amounts to 8 to 9 percent of the trash we have to send to the landfill.
- Avoid drive-through restaurants. Sitting on line with your motor running spews carbon dioxide, carbon monoxide and hydrocarbons into the air. Park your car, turn off the engine and walk in.
- Avoid single-serving packets of sugar, cream, jelly, maple syrup, ketchup, mustard, etc. Simply ask for cream in a small pitcher, ketchup in a bottle, etc.

Public Relations and Marketing

43

Make a cleaner environment part of your marketing and public relations plan.

WHY?

- Ninety-six percent of all consumers say environmental concerns have a bearing on purchase decisions, according to a poll for *Adweek's Marketing Week* magazine and the advertising firm Warwick, Baker & Fiore.

HOW!

- Read the Ten Commandments for Environmental Marketing on pages 147–150 of this book.
- Inform your customers and clients of your company's environmental activities through customer newsletters, printed on your packaging, and letters to the editor of your local newspaper.

44

Get yourself and your employees involved in environmental activities.

WHY?

- To make yourself more informed about environmental problems that exist in your community and how your company has been contributing to both the problem and the solution.
- To raise morale among your employees.

HOW!

- Start an environmental committee or group in your business or professional organization, such as the Bar Association or Lions Club.
- Get your group to work on an environmental cause such as beach cleanup or recycling venture in much the same way that the Lions Club sells light bulbs.
- Volunteer your company's services to civic organizations to work for environmental cleanup.
- Give employees time off to attend environmental organization meetings in much the same way you attend Lions or Rotary Club meetings.
- Donate money to environmental groups directly or through United Way or other group donation programs.
- Go on an "Eco-Trip" organized by Earthwatch or the Sierra Club. It will help you get an understanding of the world's environmental problems and the amount of work it will take to rectify them. These environmental service trips put you to work cleaning up natural areas or monitoring animals and

locations threatened by man's pollution. Two possibilities:

- Earthwatch, 680 Mount Auburn Street, Box 403, Watertown, MA 02172, 617-926-8200. Cost: $990 to $2,000 for two-week stints, plus transportation. Earthwatch puts you on a scientific research expedition, working as a volunteer with trained professionals around the world. Sample trips: Preserving the Great Plains by surveying grasslands, humane trapping and monitoring of black bears, humane capture and study of wild dolphins.
- Sierra Club, Outings Department, 730 Polk Street, San Francisco, CA 94109, 415-923-5630. Cost: $130 to $255, plus transportation, for trips lasting less than a week. Sierra Club will have you clearing trails and chopping brush in wilderness areas. Sample trips: Kaena Point Nature Preserve alien plant removal, in Hawaii; Bog and Oswegatchie rivers campsite restoration, Adirondack Park, New York.

45

Transform suitable company lands into wildlife habitat or donate money to land preservation groups.

WHY?

- Endangered and threatened species of wildlife and plants need protected natural areas where they can grow and prosper.
- Open spaces filled with trees and other vegetation

reduce air pollution by transforming the greenhouse gas carbon dioxide into breathable oxygen.

HOW!

- The H. B. Fuller Company of St. Paul, Minnesota, worked with the Sierra Club and Minnesota environmental officials to turn one hundred acres of their company land into a nature preserve for ponds, tall-grass prairies and upland hardwood trees.
- Ashland Oil in Russell, Kentucky, gave $50,000 to preserve the caves of eastern Kentucky.
- Amoco in Chicago gave $100,000 to preserve Illinois prairies.

For ways you can help, contact:

- The Nature Conservancy, 1815 Lynn Street, Arlington, VA 22209, 703-841-8737. They take contributions to buy endangered land.
- The Sierra Club, 730 Polk Street, San Francisco, CA 94109, 415-776-2211.
- National Wildlife Federation, 1400 16th Street, NW, Washington, DC 20036, 202-797-6800.

2

Office Recycling

Recycling office paper is a number-one priority for any business that wants to start doing its share to protect the environment. Dozens of companies have already started recycling white ledger paper, computer printout paper as well as their newspapers and magazines.

In 1989 alone, AT&T in New Jersey saved $1,383,426 in trash hauling costs by recycling their office paper. Merrill Lynch, Home Box Office, Time Warner, Consolidated Edison in New York, Ocean Spray in Massachusetts, Coca-Cola in Atlanta, Waste Management, Inc., in suburban Chicago, are all big companies that have started office paper recycling programs.

The reasons are simple. Landfill space in the United States is becoming scarce. Garbage dumps are filling up and it is becoming increasingly difficult to site new dumps or incinerators to handle the enormous amounts of trash we create. The result is that tipping fees, the amount you pay to dump your garbage at landfills, are going up. Five years ago, tipping fees were as low as $10 to $20 a truckload. Now they are approaching $150 and more.

Tipping fees for residential customers are usually paid in taxes or on a utility bill. But business usually has to pay a commercial hauler trash carting fees on a contract basis. When the hauler's fees go up, so does business's fees.

By reducing the amount of garbage your business creates, you can negotiate a lower hauling fee, or at least retard the growth of your hauling fees. This is called "cost avoidance."

Now the good news. Waste haulers are willing to pay you to take certain grades of paper off your hands. High-quality white office paper was fetching as much as $350 a ton in the spring of 1990. Computer printout paper was getting $200 on the market. The rest of your office paper, called "mixed office," was getting $10 to $20 a ton, but at least it was getting something.

What you can do as a business is separate out the office paper (there are several systems for doing this), sell that to your waste hauler and use the proceeds to offset your other trash collection costs.

THE PROBLEM AND THE POTENTIAL

The exact amounts of office paper being produced, consumed and recycled in the United States are difficult to determine. According to Matthew Costello, president of Corporate Conservation, an office paper recycling systems company in Boston, 55 percent of the trash in urban areas is commercial/industrial (i.e., business waste). Costello, who makes a living analyzing business trash, says 90 percent of that waste is dry, clean office paper, all of which can be recycled.

According to the American Paper Institute in New York, in 1986, 35.5 million tons of paper were produced in the United States. Of that total, 19.7 million tons comprised stationery, magazines, books, accounting records, computer paper and brochures.

Thirty-seven percent of high-grade office paper (stationery, letterhead, copy paper and computer printout paper) is being recovered and recycled. Thirteen percent of mixed office paper (envelopes, memo pads, colored paper, etc.) is being recovered and recycled.

Right now, most office paper that you buy that is called "recycled" is made from scrap paper collected from the printing business. This is called "postindustrial" or "preconsumer." About 95 percent of this paper is already being recycled, so there isn't much room for growth here. Two hundred of America's six hundred paper mills are now using recycled paper when they make new paper.

But new technologies are being developed so that high-grade office paper, collected from your business, can be recycled and used to make other high-grade office paper. Since only 37 percent of this paper is being recovered, there is a tremendous market for growth in this sector.

Presently, most office paper, both high grade and low grade, that is being collected from office paper recycling systems is being used to make facial and toilet tissues, napkins and hand towels. There is a huge market growing for this paper and we will talk about that later in this chapter.

In this chapter, we will discuss how to start an office paper recycling program in your business, including tips from AT&T, HBO, Waste Management, Inc., Diversified Recycling Systems in Minneapolis, Fort Howard Paper Company in Green Bay, Corporate Conservation in Boston and Sheila Millendorf & Associates in New York, one of the most experienced designers of office paper recycling programs in the country. We will discuss the problems of smaller businesses and ideas on how to buy recycled paper products. We will also describe other recycling programs, such as how to cut down on waste in cafeterias and around the coffee machine.

HOW TO START AN OFFICE PAPER RECYCLING PROGRAM

1. Get top management involved.

Without the active support of the top management in your company, you will not have a successful office paper recycling program. The company has to spend money on supplies and services to start a program and they have to dedicate staff time to organize, manage and report on the program's progress.

But the bottom line is that office recycling saves money in cost avoidance for trash hauling and it is an outstanding public and community relations tool. Any company that does not have a recycling program will be out of step with public opinion as well as the currents of business.

Finally, there are very good security and confidentiality reasons why you should recycle your office paper. If you just throw your office paper away, information about your employee records, payroll, taxes, costs and profits, client lists, billing and corporate planning could very well be floating around public garbage dumps for all the world to see.

Office paper that is recycled is baled and wrapped in steel bands, loaded on trucks and taken to paper mills. The public thus has no access to your confidential documents.

2. Do a survey and feasibility study.

How many employees do you have? The rule of thumb is that one employee creates one pound of recyclable office paper per day. Vendors who buy recycled office paper talk in tons, not pounds. That means you have to generate enough recyclable paper to make it worthwhile for a hauler to make stops at your address.

Another rule of thumb is that you need a minimum of 150 to 200 employees to launch and manage a recycling program all by yourself. If you don't have enough employees, you will probably need to talk to other business tenants in your building or your office park to come up with enough paper flow to start a program.

What types of paper do you produce? There are eighty-one different types of paper and all of it can be recycled. AT&T recycles everything. Their motto is "If it tears, recycle it." Other companies recycle only the high-grade office paper and computer printout paper. The demand for your office paper in your area and the guidelines established by your vendors will determine what you do and don't recycle.

Most buildings are not designed to store a lot of wastepaper. You will have to analyze the storage areas in your building and on the loading dock to see what sort of problems you will encounter. But if you are going to save a lot of money by recycling, it will be worth your while to rearrange existing space or create new space. All it has to be is clean, dry and accessible.

3. Find a vendor.

The type of vendor you get will determine what kinds of paper you can and will recycle. Some vendors are only willing to buy your high-grade office and computer paper because it brings them the highest price on the paper market. Others want to buy all of your office papers because they have the technology to utilize it all to make new paper products. Here's what to do:

- Call your state, county or city office of recycling or solid waste. It could be located in their office of environmental protection, or department of health, public works, conservation. Ask them to give you a list of recycling vendors, specifically, people who are in the business of collecting office paper to be recycled.
- Look in the telephone Yellow Pages, either business or consumer Yellow Pages, under Recycling, Waste or Waste Paper.
- Call some of the major companies in your state or region and ask them which vendors they use. Create a list of at least six to twelve vendors whom you can call.
- Call the vendors and ask them what types of paper they will pick up, how they want it separated and bundled, how often they will come by and how much paper you need to generate to get them to make a pickup. Ask them how much they will pay you for what types of paper you generate. Ask for references and check them out. (NOTE: You may not find a vendor who is willing to make a stop at your

office if you do not generate enough paper. In this case, you will have to cooperate with other offices in your building or simply take your office paper to a municipal or community recycling center that accepts office paper. The best solution for small business is mandatory city-sponsored curbside pickup and recycling of office paper.)

- Ask the vendor what their security program is to keep all your papers confidential. You want to use a vendor who is bonded and will give you a certificate of destruction, which proves they will use security measures to protect your information.
- Make an on-site visit. It is too easy for waste haulers to promise you security services, but then not back them up in practice.
- Choose a vendor who can grow with you. You want one that might be willing to take a smaller amount of paper from you in the early stages of your program, but is then capable of taking larger amounts later on.
- Choose a vendor who will be willing to take different grades of paper. You may start out collecting only computer paper or white ledger, but you might want to expand your paper collection process later on.

4. Devise a system that works for your business.

All office paper recycling systems start with the individual employees being responsible for sorting and recycling the office papers that they generate. It is important that the employees get personally involved so that they can feel a sense of accomplishment, a sense of protecting the environment, a sense of helping make their company grow and prosper.

There seem to be two main schools of thought on

several key points in a recycling program, and your choice on each one will determine how you should lay out your system. Ask yourself these questions:

- Will we recycle all office paper or just the higher grades that fetch the higher prices?
- Will we put the recycling bins on top of the desk or under the desk?
- Will the employees empty their individual recycled paper into central collection points or do we leave that to the janitorial staff?

The type of office paper recycling bin or folder you use depends on what types of paper you are going to recycle. If you are going to recycle all office paper, you will need an employee bin large enough to handle all the paper. If you are going to recycle only high grades of paper, you can use a smaller bin or folder. If you are going to separate high-grade from low-grade and recycle both, you will need a two-slotted bin for that. A two-bin system is as much as you will want to incorporate, one for trash and one for recycled papers.

Sheila Millendorf, president of Sheila Millendorf and Associates in Rockville Center, New York, has been designing office paper recycling systems since 1979. She founded the Office Paper Recycling Program for the Council on the Environment, a quasi-governmental group attached to the mayor's office in New York. She designed systems for AT&T in Manhattan, Chemical Bank, Consolidated Edison, Merrill Lynch and dozens of other places in and around New York.

She is a strong believer in the desktop folder for recycled paper. She wants the recycling bin to be strategically placed on top of the employee's desk. She discourages companies from putting a recycling bin next to the trash can on the floor next to the employee's desk because she wants people to know that office paper to be recycled is an asset and it deserves to be placed on top of the desk.

She says a desktop system prevents contamination of the recyclable paper because some people will still carelessly throw trash like coffee cups, food wrappers and bubble gum into a recycling bin if it is on the floor. For her, a desktop system, usually a flexible plastic folder, is more convenient to use and it is easier for the employee to carry a folder of the recyclable papers to a central recycling receptacle than to carry a larger cardboard or hard plastic bin. Finally, a dual waste basket system takes up too much room around the employee's desk area.

Home Box Office and many other of Mellendorf's clients use the desktop system and they are very happy with it.

Cheryl LaPerna, the recycling coordinator for AT&T in New Jersey, was also a believer in desktop bins when she started their intensive program in 1984. But, four years later, she was disappointed that their recycling rate was only 50 percent successful, generating 12.5 tons of office paper a month for recycling.

She switched to under-the-desk trash pails, one for trash and one for paper to be recycled, and their recovery rate shot up to 90 percent and they began generating twenty-five tons of paper a month.

Waste Management, Inc., in Oak Brook, Illinois, the world's largest trash hauler, uses a dual basket system at its corporate headquarters. New employees are assigned a desk, with their own paper recycling receptacle and given an orientation on office paper recycling. The recycling bin is white and has the company's recycling logo on it. The trash can is black, but any two distinct colors can be used to differentiate between the recycling bins and the trash bins.

Once again, it is important that the individual employees do the source separation themselves. But who takes it from there is a different story. Some systems prefer that the employees themselves take their recycling bins to a larger, centrally located recycling bin. Some prefer that the janitorial staff empty all bins at the end of the day.

Everybody agrees that large recycling bins be located next to or near the office copy machine. That is where a large portion of your office paper is being generated. The janitorial staff should definitely take over the operation from there and cart the wastepaper to a large storage area near your loading docks or back door.

5. Order supplies.

Obviously, you will need to buy some additional office equipment to start a program. First contact your existing office supply company and ask them what they have in stock or what they can easily order for you in the way of recycling bins.

According to Tom Lewis of the Johnston Paper Company in Auburn, New York, it will cost approximately $200 to supply an office of ten people with equipment to recycle office paper.

Rubbermaid hard plastic wastebaskets or sturdy cardboard boxes from the Quill catalogue can cost as little as $3 to $4 each.

Specially designed desktop recycling bins can be purchased for $6.95 each or three for $18.95 from the Seventh Generation catalogue. Your existing office products supplier will have similar products at similar prices. Write or call Seventh Generation, Colchester, VT 05446-1672, 800-456-1177, for a catalogue and to place orders. This is a very good place for small businesses to buy just a few recycling containers to get your program off the ground at a very low cost.

Another good source is Diversified Recycling Systems, 5606 North Highway 169, New Hope, MN 55428-3099, 612-536-6662. They have dozens of different types of office paper recycling bins and containers made of plastic, cardboard and steel. They also have the larger bins your janitorial staff will need to collect and transfer the materials from the desk to the storage areas. Most office paper and supply vendors will either carry items or be able to order from Diversified. For larger wholesale orders, you can order from them directly.

6. Develop an information and education campaign.

You have to have a fully developed information, orientation and educational campaign in place for your system to work, once it has been devised and put in place.

An information and education program can be as simple or as complex as you want to make it. If you are a small business with fewer than twenty employees, you might simply call an office meeting during lunch hour or first thing in the morning over coffee in the conference room. If you have more employees, you might break up into groups and conduct several orientations.

Orientations can be as simple as showing employees which types of paper to recycle and which box to put them in. You can buy stickers or make them up; usually the company that provides you with bins will have stickers, reminding employees of what goes where. New employees can have recycling as part of their general office orientation.

Waste Management, Inc., started the process one month before they instituted their program by sending each employee a letter, signed by top management, telling them about the soon-to-be-implemented program. They sent them another letter a week later, another letter another week later and, finally, on the day the program was put into effect, all employees went through an orientation procedure in the company auditorium or cafeteria that included videos and slide-tape presentations.

They have also appointed a group of "office paper recycling coordinators," one on each floor of their headquarters building, ten in all working with eleven hundred employees. These OPR coordinators keep an eye open to be sure employees are using their recycling bins properly, if any bins are damaged and need to be replaced and generally to answer any questions about recycling from employees.

AT&T has a very active informational program that features brightly colored posters strategically placed around the office. They constantly monitor their paper collection program and keep employees informed about how many trees they have saved from being used for paper and how much money the company has saved through their efforts.

Most, if not all, systems have stickers attached to their individual recycling bins telling employees exactly what can and cannot be recycled.

Here is a checklist of things you will need for an educational program:

- Letters or memos for each employee telling them about the program. The letters should explain how the system is going to work, how the environment and the company will benefit from the program and invite comments or constructive criticism from employees.

- Brightly colored posters, signage and stickers on receptacles and around the office reminding staff of how the system works and where the proper receptacles are located.
- Follow-up. Include information about the program's progress in the company newsletter or other information organ.
- AT&T discourages competition between departments on recycling rates because people can, at worst, behave in an underhanded manner and, at best, be distracted by competition. Waste Management, Inc., plans to make a game out of their program by rewarding the entire staff with free doughnuts or sandwiches if the whole company meets its recycling goals.

7. Buy recycled office products.

It may have been true twenty years ago, but today's new office paper made from recycled paper is no longer rough and inferior and it does not jam up copiers and other office machines.

According to Matthew Costello of Corporate Conservation in Boston, clients to whom he has shown recycled office paper and virgin office paper cannot tell the difference.

Most recycled office paper that is on the market today is made from scrap paper collected from the paper and printing industries. This has been going on for years and it is a wonderful thing. But it does not do anything to help cut down on the amount of solid waste that is generated and sent to our already overburdened landfills.

It is very important that postconsumer office paper, which means office paper that has been used once in an office, be recycled and made into new office products.

Now, some office paper recycling programs, like AT&T's, are successful at cutting both costs and waste, because they recycle all their grades of office paper, even envelopes with plastic windows and gummed labels. They sell their paper to the Fort Howard Paper Company of Green Bay, Wisconsin, because Fort Howard has the technology to make new paper products out of mixed office paper.

Fort Howard has been recycling office paper for fifty years and yet they aren't even in the lumber and forestry business like other paper companies.

They take mixed office paper and make it into napkins, hand towels, facial tissue and toilet paper. Furthermore, they don't add chemical inks or dyes, colorings or scents to their recycled paper products and they use a bleaching process that does not give off any dioxin. Their consumer line of recycled paper is called Green Forest and their business line is called Envision.

Fort Howard and other paper companies are also working to develop technologies that take office paper and recycle it into more office paper.

What you need to do is buy recycled paper products whenever you can. Sometimes the price is competitive with virgin paper and sometimes it is a little more. In mid-1990, Quill was selling their blue roll computer paper, made from recycled newspapers, for less than virgin paper. The New York City Department of Sanitation has instructed their purchasing department that they are willing to pay up to 10 percent more for recycled paper than for virgin. Demand for recycled paper will stimulate supply, economies of scale will be enacted and prices will go down.

The best way to help this is to make buying recycled paper products, and postconsumer paper products when available, company policy. Instruct your purchasing department to specify recycled paper products when ordering from your existing suppliers. The products are out there and all they have to do is look for them.

8. Other office recycling programs.

Home Box Office has a very active recycling program in their New York City headquarters. All employees are issued a reusable ceramic coffee mug so they no longer have to use polystyrene or paper cups for coffee. The idea was to cut down on the amount of solid waste they generated, but they found that their coffee costs have dropped dramatically.

The reason? They found that employees discontinued their habit of grabbing a clean disposable cup and filling it with coffee each time they walked by the coffee machine. Too often, the cup was left on the desk and the coffee went cold. They would toss out the cold coffee and cup and pick up a new one next time they went by the machine. HBO found that if each employee is made responsible for washing and maintaining his or her own HBO mug, they stopped wasting so much coffee.

HBO's cafeteria also recycles all its cans, bottles and jugs. They don't use any throwaway trays or plates. All items are washable and reusable. They have some carry-out packaging, but very little.

No one has yet done a complete cradle-to-grave cost analysis of disposables versus the dishwasher and detergent costs of reusables. But schools generally prefer disposables because they have significantly higher labor costs if they use reusables in their cafeterias. But if accelerating landfill costs are factored in, the cost of throwaways and reusables will probably be roughly equal.

CHAPTER SUMMARY

1. Start an office paper recycling program. Get top management and employees actively involved.
2. An office paper recycling program saves our environment and saves the company money on disposal costs.
3. Buy recycled office paper products, from hand towels to computer paper. Buy postconsumer products if possible.
4. Recycled paper comes in many grades, too. You can guy good-quality recycled paper that does not cause jamming problems for copy machines and other business equipment like envelope stuffers or billing machines.

3

Dealing with Indoor Air Pollution and the Sick Office Syndrome

In the 1970s, two trends developed that had a profound effect on the quality of air we breathe, and therefore our health, while working indoors. The first trend was the energy crisis that forced us to tighten and weatherize our buildings. This reduced energy costs, but it also reduced the free flow of fresh air.

The second trend was the increased use of synthetic building materials plus solvents and other organic compounds around the office. The combination of these two trends, it turns out, has caused us to start breathing air contaminated with chemical gases and biological particles.

First, the high cost of fuel oil, natural gas and electricity needed to heat and cool our buildings made us look for ways to eliminate waste. Energy conservation became a dollars and cents battle cry. We weatherized our buildings, sealed windows and doors against drafts, and modified ventilation systems to keep our energy budgets in line. The result: indoor air was not being refreshed often enough with fresh outdoor air to dilute the bacteria, chemical vapors or tobacco smoke present in the indoor environment. Secondly, a whole new array of money-saving building materials was introduced. Many of these new products were synthetic, such as carpeting, wall dividers, insulation, ceiling tiles, pressed woods for under flooring and desk- and tabletops.

Against this background of energy-efficient buildings with restricted fresh air flow and the introduction of new building materials, people who work in offices have started to complain of headaches, dizziness, fatigue, cold and flulike symptoms, sore throats and other minor illnesses.

A new term was developed, "sick office syndrome." Not a lot is known about this problem yet and physicians are reluctant to qualify it as a full-fledged disease. But that doesn't mean this problem is not costing business plenty. The Environmental Protection Agency puts the annual cost to business in lost productivity, extra sick days and direct medical costs at over $10 billion a year for sick office syndrome.

The outbreak of Legionnaires' disease in 1976 that left twenty-nine people dead in a Philadelphia hotel has been attributed to pneumonia-causing bacteria that grew in the building's ventilation system and which were then circulated around the building. Legionnaires' disease continues to spread (twenty-six cases of Legionnaires' disease were reported in a South Dakota hospital from 1985 to 1988. Ten people died) and is one of the worst-case scenarios of sick office syndrome.

Furthermore, employees are starting to sue their companies over illnesses they have contracted while working in the office. The *Wall Street Journal* says over a dozen indoor air pollution suits have been filed in the last two years and more are expected.

Cleaning up indoor office air pollution does not have to be an expensive operation. A properly cleaned ventilation system with increased fresh air flow is often the cure. There are dozens of environmental consultants and industrial hygienists who are specializing in office air pollution. Architects and the heating and cooling industry are beginning to take notice of the problem.

It is important to bear in mind that just because your office has asbestos in the ceiling tiles or formaldehyde in the desktops, that does not mean that your office air is contaminated. According to the National Institute for Occupational Safety and Health, inadequate ventilation, or simply breathing stale air, is the cause of over 50 percent of the reported cases of sick office syndrome.

This chapter will explain what the causes and sources of indoor air pollution are, how you can detect them without hiring outside help and some cost-effective methods and techniques that can help you clean up your workplace and enhance the productivity of your employees.

SOURCES OF INDOOR AIR POLLUTION

Asbestos—Asbestos is a naturally occurring group of mineral fibers that is used in thousands of products. Office buildings are likely to have asbestos in pipe and duct insulation and ceiling panels. Asbestos only becomes a problem when the building material it is combined with starts to deteriorate or if the material is disturbed during remodeling, reconstruction or removal. Then the tiny fibers can get airborne and enter people's lungs.

An estimated 20 percent of commercial buildings in the United States contain asbestos, according to the National Asbestos Council of Atlanta. Nearly ten thousand people will die in 1990 from asbestos diseases related to occupational exposure.

The short-term health effects of inhaling asbestos fibers are relatively benign. But the long-term effects of asbestos in the lungs include scarring, called asbestosis, and cancer. These adverse health effects may not show up until fifteen to forty years after the employee is exposed to asbestos fibers.

Most asbestos-related illnesses have occurred in workers in the asbestos industries. But since there is no known safe level of exposure to asbestos, it is best to take precautions.

SOLUTIONS: If you find asbestos in your building and it is in good shape and undamaged, it is best to leave it alone. But, according to the New York City Environmental Protection Commission, 87 percent of asbestos found in buildings *is* damaged. Once asbestos is detected, it is wise to have it inspected by a professional. If you determine that the asbestos should be removed, have it done by a licensed professional.

Asbestos can be covered and sealed before it is damaged to protect it, but once it is crumbling and exposed, it is best to have it removed.

Formaldehyde—Formaldehyde is a common chemical compound that is used extensively as an adhesive in building products. Particle board for subflooring, paneling, cabinets and furniture is held together with compounds containing formaldehyde. Some carpeting, drapes and upholstery also contain formaldehyde. Cigarette smoke is also a source of formaldehyde.

The problem with formaldehyde is the "off-gassing," that is, the vapors that emanate from products made with formaldehyde. These vapors can contribute to nausea, dizziness, eye, nose and throat irritation, rashes, coughing, fatigue and allergic reactions.

SOLUTIONS: Consult with your heating and ventilation contractor to increase airflow in your office and to reduce humidity levels if they are too high. Correct humidity and adequate airflow usually will remove formaldehyde gases.

When you are planning to remodel your office, try to use natural wood or metal for desks, bookshelves, trim, doors, tabletops, etc. Instead of particle board for subflooring, try to use exterior-grade plywood, which is not made with formaldehyde.

If after consultation with a professional you find that you have an unusually high level of formaldehyde and ventilation and humidity adjustments do not abate the problem, you will have to remove the sources or cover them with vapor barrier paint or wallpaper coverings.

Biological—These are bacteria, molds, pollens, fungi and viruses that are common in the air everywhere. The problem arises in that these microscopic plants and animals thrive in modern energy-efficient buildings that seem to have a shortage of fresh air. They also like standing water and pools of water on carpets caused by a poor humidification system in buildings.

The health effects of biological problems include the spreading of flu, chicken pox and measles, as well as allergic reactions, rashes, watery eyes, sneezing, fatigue, coughing and respiratory and digestive problems. Legionnaires' disease and "humidifier fever" are two health problems caused by bacteria, molds, fungi, etc., spreading through a poorly maintained ventilation system.

SOLUTIONS: Find and clean up the sources where biological dangers can breed, such as wet or damp carpeting, dirty washrooms and uncleaned air filters in humidifiers and ventilation systems.

Volatile Organic Gases—VOGs are produced by the synthetic chemical industry and come from a variety of sources in an office. Some sources of organic gases are: furniture, paints, strippers, dry-cleaned clothing, bathroom air fresheners, adhesives, glues, chemicals used in office copiers, signature machines and blueprint copiers, felt tip markers and pens, correction fluids, janitorial cleaning products, telephone cable, room dividers, carpets and artist's supplies.

The EPA has found as many as five hundred organic compounds present in indoor office air, including benzene, chloroform, methyl alcohol, butyl methacrylate and trichloroethylene.

SOLUTIONS: Ventilation is the key to controlling organic gases. Once again, insure that overall ventilation levels are high enough to bring in enough fresh air to dilute ordinary VOG levels. In areas such as copy and duplicating rooms, graphic arts departments and even the mail room if adhesives are used there, extra ventilation, including direct outdoor ventilation, may have to be installed.

Eye and respiratory irritation, fatigue, coughing, rashes and allergic reactions are some of the short-term health effects. Long-term effects are less clear, but some VOGs have been linked to cancer.

Carbon Monoxide—Carbon monoxide is a deadly gas that is emitted from automobile and truck exhaust as a result of incomplete combustion. It usually enters office buildings by way of intake vents that are too close to garages and loading docks. Buildings located in congested urban areas or close to heavily trafficked roadways are particularly susceptible to carbon monoxide getting into their indoor air. Other sources of carbon monoxide air pollution are from cafeterias and restaurants on other floors of your building that are improperly venting their gas cooking ranges and ovens.

Health problems associated with carbon monoxide include fatigue, headache, dizziness, disorientation and respiratory problems.

SOLUTIONS: This may be one of the easiest problems to detect. Look to see where your air intake vents are located. If they are too close to loading docks or garages, move the intake vents to another location where clean, fresh air can be accessed.

Tobacco Smoke—Tobacco smoke is emitted from smoking tobacco products in pipes, cigarettes and cigars. Tobacco smoke contains over 4,700 different compounds, including formaldehyde, carbon monoxide, nitrogen dioxide and breathable particles.

The problem with tobacco smoke is that the smoker is forcing his or her fellow workers to inhale secondhand smoke, now being called "environmental smoke."

People who work in the same office space with smokers are called "passive smokers" and new studies are indicating that their health risks of getting heart disease or cancer are significantly higher than if they were not exposed to environmental tobacco smoke. The American Lung Association reports that over one-half of nonsmoking office workers were acutely affected by environmental smoke. It is estimated that over half of all U.S. businesses allow unrestricted smoking in the office.

Health problems associated with tobacco smoke include eye, nose and throat irritation, coughing and headaches. Lung cancer and heart disease are also now being connected to environmental tobacco smoke.

SOLUTIONS: Eliminate all tobacco smoking in your business environment. Encourage your employees to quit smoking and provide incentives. Provide heavily ventilated designated smoking areas away from other workers. Insist that your smoking employees leave the building when they smoke.

Pesticides—The EPA now says that indoor air is more polluted and hazardous to breathe than outdoor air. The main cause of that is pesticides, including orthophenylphenol—an active ingredient in many Lysol-brand disinfectant products—chlordane—a widely used termite killer until it was banned in 1988—and others.

You may be in a high-rise building that seems far away from any bug infestation. But food attracts bugs, and if there are food service establishments in your building, such as restaurants, caterers, executive dining rooms, cooking schools or a pantry with a coffee pot and refrigerator for employee lunches, you may be exposed to a pesticide spraying program. It is even possible that your office air can become contaminated with pesticides or herbicides if the grounds around your building are treated with chemicals and the vapors are sucked into the building's ventilation system.

SOLUTIONS: Make sure that any pesticide spraying is conducted on weekends and that the building is thoroughly ventilated before employees return to work on Monday. Ask your exterminator or staff person responsible for the pesticide program to review with you their methods and types of chemicals they are using.

It is important that the mixing of pesticides be done outside in a well-ventilated area, that they are using only recommended doses of the chemical and that they are using the lowest doses possible in their eradication program. If you are experiencing only a minor infestation, you might want to switch to less hazardous methods, such as traps.

Lead—Lead in paint on walls and woodwork is a serious source of indoor air pollution. Although lead was banned from paint in 1977, the U.S. Public Health Service reports that 3 million tons of lead still exist in old paint on walls around the country. It is unlikely that any building constructed after 1977 was painted with materials containing lead, but the possibility still exists, as lead-enhanced materials were used up after the ban.

In any event, lead is not a problem until it is disturbed by sanding, chipping or other types of remodeling. That is when the lead in the paint gets released into the air in the dust created in remodeling.

The health effects of lead poisoning are not so acute for adults, although there are cases of "yuppie lead poisoning"—families who have picked up high levels of lead poisoning while doing restoration on older homes.

Children under six years of age are the most at-risk because lead poisoning attacks their developing neurological systems and can retard their brain and motor skills. In other words, lead poisoning can lessen or retard a child's intelligence levels. According to Karen Florini, an attorney for the Environmental Defense Fund, and Dr. Ellen Silbergeld, a toxicologist, co-authors of a 1990 study on lead poisoning, children can become exposed to lead poisoning simply by inhaling or ingesting dust particles their parents bring home on their clothes from an office in an older building that is undergoing remodeling or restoration.

The U.S. Centers for Disease Control are likely to lower the guidelines on the amount of lead present in the bloodstream necessary to constitute lead poisoning. This will be the second time they have lowered this level in ten years. They now believe that lead poisoning is more pervasive than previously believed.

SOLUTIONS: If your business is in an older, pre-1977 building and you suspect that the walls or woodwork is covered in lead paint, have it tested. Contact your local or state board of health for advice on how to get it tested. If you do have lead paint, and you want to continue with remodeling, you must have the lead removed by experts. Lead abatement is just as serious a problem as asbestos abatement.

There is not yet a body of court cases testing the liability of lead poisoning contracted in an office atmosphere. The potential for liability does exist, most likely with the owner of the building, but it is also possible that an employer might become involved in any liability suit. If you are planning on conducting restorations, keep your employees isolated and protected from the dust and particles from the work area.

BLUEPRINT FOR HEALTHY INDOOR WORKPLACE AIR

Now that we have discussed the sources and causes of indoor air pollution, let's move on to abatement. This is a three-step process.

1. Survey employees for reported health problems, if any.
2. Next, conduct a workplace evaluation that includes documenting the history of the building, recent renovations and ventilation maintenance, plus an on-site walk-through to look for problem areas.
3. Apply solutions such as ventilation modifications, changing office design and fabrics, installing air-cleaning plants.

EMPLOYEE HEALTH PROBLEM SURVEY

If you are the employer, and one or more of your employees complain about health problems that occur on the job and not at home, you will want to start the process of determining what the problem is and how you can solve it.

First, question all of your employees about possible health problems. Some questions you will want to ask include:

- What sorts of health problems are you experiencing: irritations of the eyes, nose or throat, coughing, nausea, fatigue, dizziness, sneezing, unusual stress, grogginess, headaches or any other?

- Have you ever reported these symptoms before at another place of employment? Have you spoken with a doctor about them? Has the doctor made a diagnosis or are you now being treated?
- Are there certain places in the workplace where the symptoms occur?
- Are there certain times of the day, week or month when they occur?
- Do the symptoms begin when you come to work and lessen when you leave?

At this point, if you feel the building may be causing some of the symptoms, you will want to continue with the building survey. Ask yourself these questions:

- When was the building constructed?
- What materials were used, including insulation, flooring, paints, ceilings, cables, carpeting, partitions, windows, etc.?
- What was the original purpose of the building? Office, manufacturing, food preparation, etc.? What is it being used for now? In other words, if the building was designed for offices, and a print shop, dry cleaners, photo lab or other shop was added later, have there been modifications in the ventilation system? Remember, fumes from one workplace can be transported to another.
- Has vehicular traffic in the neighborhood significantly increased since the building was designed?
- Do the windows open? Were they designed to open? Have they been sealed shut?
- Has the building been recently weatherized or insulated?
- Has the building recently been renovated, repainted, wallpapered? Have new phones, computer systems or office partitions been installed?

- Do you have central air conditioning? Do window air conditioning units drip water on carpets?
- Are the bathrooms kept clean and are they ventilated?
- Do you have an unusually large printing, photocopy or graphics department that smells of chemicals when you walk in the room?
- Do you permit unrestricted tobacco smoking in the workplace?

If you have employees who are complaining of the symptoms of sick office syndrome and you have noticed several problems with the design of your building or office, you are probably ready to call in some professional help.

First, talk with building maintenance. They may be aware of the problem and have already started to correct it, especially a well-maintained or modified ventilation system.

If you need more help, contact a heating, air conditioning and ventilation professional. Look in the phone book. Interview several and check their references before you hire them.

Finally, there is a new breed of environmental testing expert who will come to your building and test the air for smoke, carbon monoxide, chemical or biological levels, inspect the ventilation system and give you advice on what you can do. Look in the Yellow Pages under Air Pollution Control, Air Pollution Measuring Service, Consulting Engineers or Industrial Hygiene Consultants.

Environmental testing for sick office syndrome is a very new concept and there may be fraudulent operators in the market. There is a lot of money to be made in this growing field. It is estimated that $1.4 billion will be spent on environmental testing in

1990. If you are considering hiring a service to test the air in your workplace, check this list:

- Ask your local or state departments of health or environmental protection for a list of companies that they can recommend. Your local or state American Lung Association may also be able to recommend services.
- Ask your colleagues in business and professional organizations for names of services that they have used.
- Ask the services you call if they are certified to conduct tests for any government agencies and for which contaminants.
- Ask the services if they have ever been decertified by a state agency to conduct tests.
- Ask the services for references and check them out thoroughly before you hire them.

4

Energy Conservation and Efficiency: How Business Can Prevent Air Pollution and Save Money

According to the twentieth annual report of the President's Council on Environmental Quality, "U.S. consumption of energy is linked to many of this country's most pressing environmental concerns. Emissions released to the atmosphere from fossil fuel combustion contribute to atmospheric degradation. Sulfur and nitrogen emissions contribute to acid rain. Nuclear fuel, while causing no acidic or CO_2 emissions, produces dangerous radioactive wastes."

The report goes on to say that "improvements in energy efficiency—getting the same amount or more energy services for less energy input—has tremendous potential for preventing pollution."

Energy conservation and energy efficiency, which is now being called "demand side control," is another positive step that business can take to prevent the amount of pollution it creates. Demand side control is primarily a management function, because in most cases the technology is readily available and workplace models have been in existence for twenty years.

During the energy crisis of the 1970s, many business people and government officials, especially those in the coal- and oil-rich states, believed that producing more energy was the solution to our energy woes. Production did increase, nuclear power got a big shot in the arm, even solar and renewable energy sources started to become cost competitive. But it was conservation, brought about by price increases and weatherization retrofits, that broke the back of the OPEC cartel and finally stabilized the energy market.

During the Reagan years of low fuel prices, energy conservation lost its urgency and luster. Air quality improvements were significant during the early 1980s. Now, although things have taken a turn for the worse, fears about acid rain, global warming and deteriorating air quality have prompted people to start taking a renewed interest in energy consumption and the environment.

Chapter 5 will take a closer look at environmental problems caused by burning fossil fuels for transportation needs. This chapter will take a look at the environmental and health problems caused by burning fossil fuels, primarily coal and fuel oil, to produce electricity to heat, cool, ventilate and light our businesses and offices, and electricity needed to power our manufacturing processes.

We will look at what's new in solar power and how a New England utility system has teamed up with an environmental group to offer business customers new ways to save energy, save money and protect the environment.

First we will explain the concept of energy efficiency, then present a discussion with Richard Aspenson, a mechanical engineer and founder of 3M's innovative and productive twenty-five-year-old energy conservation office. Aspenson says energy conservation is a profit center.

Then, we will take you on a tour of your own workplace and show you where and how you can make significant air pollution improvements in lighting, heating and cooling, ventilation, fans and motors.

EFFICIENCY

Energy efficiency is not exactly a fuel source, but that is how scientists, utilities and government regulators are looking at it now.

Amory Lovins, the somewhat controversial economist and founder of the Rocky Mountain Institute in Snowmass, Colorado, estimates that energy efficiency could prevent enough pollution to solve the global warming problem. More importantly, Lovins says that a profit of $1 trillion a year could be saved by the American economy each year through energy efficiency.

Lovins calculates that one-fourth of U.S. electricity output could be saved with more efficient lighting, another fourth with efficient electric motors and one-fourth in more efficient appliances and other equipment. He suggests screwing in a more efficient light bulb as an alternative to planting a tree to reduce and prevent global warming.

But Lovins is not a voice in the wilderness. The U.S. Department of Energy has included energy efficiency as a key component of their new comprehensive energy strategy for the United States. The DOE is working with states and utilities to revise rate regulations to encourage energy conservation.

The President's Council on Environmental Quality's twentieth annual report states that "opportunities to advance pollution prevention through energy conservation are still plentiful," and "opportunities for utilities to conserve energy will expand if there are incentives to invest in demandside management programs." Demand-side management is translated as energy conservation and effficiency.

The report states that "one high efficiency light bulb, which lasts ten times longer than low efficiency alternatives, can eliminate the combustion of 220 to 382 pounds of carbon," which is a suspected cause of global warming.

ENERGY EFFICIENCY: A CASE STUDY

Nearly one-third of all U.S. carbon dioxide emissions, the greenhouse gas, come from electric utilities. The solution to pollution for electric utilities and business is to work together to improve efficiency.

Electric rates traditionally have been based on a utility's projected needs to meet consumer demands by building more capacity. Even though electricity consumption has been growing steadily over the past twenty years, the costs of building new power plants, both environmental and construction costs, have skyrocketed. Government regulators have always given utilities their rate of return based on new building costs, overlooking conservation investments as a way to grant a fair rate of return.

Utilities had started energy conservation programs during the energy crisis. But most of that was aimed at residential customers in the form of energy audits, rebates, loan programs and installation of water heater blankets and flow-restricted shower heads.

But a new program in New England is aimed at business, commercial and industrial customers. The New England Electric System, which provides power in Massachusetts, Rhode Island and New Hampshire, working with the Conservation Law Foundation in Boston, has embarked on an aggressive commercial energy conservation program.

The utility wants to invest $65 million in conservation efforts, such as paying consumers to buy more energy-efficient light bulbs and lighting systems as well as more modern efficient appliances and heating, cooling and ventilation systems.

The key is that they want government regulators to grant them a rate of return on investment equal to or greater than the rate of return on traditional new building investment. This initiative is based on three principles:

1. Energy efficiency is nonpolluting and environmentally clean.
2. It is cheaper than building new capacity.
3. It is quicker than building new generating stations.

CASE STUDY: THE ENERGY AND ENVIRONMENT EQUATION AT 3M

But business does not have to wait for government regulators and the utility industry to start making real contributions to air quality and their own bottom line. The Minnesota-based chemical giant 3M estimates that since they started their energy-efficiency program in 1973, they cut energy use by 55 percent and saved the company $732 million through 1989. This energy savings, accomplished through increased efficiency, added 40 cents a share before tax on U.S. earnings and 24 cents a share after tax. That is a bottom line anybody can understand.

3M's energy-efficiency program has also helped clean up the environment. In 1988, their program reduced carbon dioxide emissions by 3.35 billion pounds, sulfur dioxide emissions by 13.6 million pounds, carbon monoxide emissions by 2.6 million pounds and nitrogen oxides by 5.6 million pounds.

Energy, Jobs and Profit at 3M: A Discussion with 3M's Energy Office Director Richard L. Aspenson

"Energy conservation has been a way of life for me," says Dick Aspenson, who started 3M's energy conservation program in the late 1960s. Besides working for 3M, Aspenson, a mechanical engineer, is active on the energy committee of the National Association of Manufacturers and has participated in U.S. Department of Energy studies and information gathering.

"The program became formalized in about 1975, right at the height of the energy crisis. When we set up, we had three goals in mind:

"1. Maintain a reliable energy supply for the company.

"2. Have a direct impact on the cost of manufacturing by reducing energy consumption.

"3. Make sure the huge wave of new plants 3M was building was designed as energy efficiently as possible.

"The first thing we did was set up new design standards for energy efficiency in a plant retrofit we were planning. Next, we ran preliminary energy audits on all the facilities to identify opportunities for energy conservation. We matched up our design standards with the audits, learned what could and couldn't be done and modified our standards.

"Our first step was to implement the no- or low-cost energy saving recommendations. Things like turning off equipment, lights, fans and motors when not in use, turning down thermostats in winter and turning them up in summer. This was a management directive and our people suffered a little bit, but it was worth it.

"We showed our people that they could have an impact on the cost and use of energy. We called our campaign 'Energy, Jobs, Profits,' because we figured everybody could relate to at least one of those ideas.

"Since day one our department has been a profit center for 3M. Energy conservation generates profits and the people in top management saw that and supported us.

"After the easy stuff I just mentioned, we worked on comprehensive plant audits to find opportunities that would take some investment, but would be profitable in the long run. We calculated the cost and savings, projected profit and loss and began to market our plans to the respective divisions. We didn't have the authority to enforce these new plans, so we had to make presentations to management.

"It was over these years that we cut energy use by 55 percent and saved the company nearly $740 million.

"But, then, during the 1980s, energy conservation and use became a low-profile issue. It has been hard to motivate people to save up until now. I think concerns for a cleaner environment are going to be very powerful in the 1990s. We are calling our new theme 'Energy, Environment and Economics.'

"Business should become very proactive in energy conservation. Every utility and business should be required to come up with a demand-side control program. For public relations purposes, as well as profits, they should start an energy conservation program immediately. Corporations are hiring environmental officers, and they should remember that energy use is a prime contributor to air pollution."

HOW TO CONDUCT A DEMAND-SIDE CONTROL PROGRAM IN YOUR BUSINESS AND CLEAN UP THE ENVIRONMENT

1. Get top management involved. Appoint a ranking corporate officer to oversee the project.
2. If you rent your space, get the support of your landlord or management company.
3. Do an energy audit. You may want to bring in a professional. Contact your utility for their auditors or ones they respect and recommend.
4. Some professional energy auditors can give you a complete computerized breakdown of where your energy losses are, what it will cost to fix them and how fast the payback period is.

PLACES TO LOOK FOR ENERGY CONSERVATION SAVINGS

Lighting

More than 20 percent of all electricity generated in the United States each year is used for lighting. As much as 30 percent of the electricity used in the commercial sector is used for lighting. Cost savings of 20 percent to 50 percent are not uncommon.

- Use lower-wattage bulbs in fixtures that are supplying too much lighting. Too much lighting can lead to glare and reduced productivity. Certain areas of your business may not need all the lighting you are giving them.
- Remove unnecessary lights in fixtures. Be sure to unhook the ballasts that supply that bulb with electricity because they keep using electricity even if the bulb is removed.
- Install high-efficiency fluorescent lamps, replacing standard ones.
- Set timers on lighting to reduce lighting after hours or simply be sure to turn lamps off when not in use.
- In any new building design or remodeling, be sure to take advantage of natural lighting from windows and skylights. Con Edison in New York gave the Natural Resources Defense Council in New York a grant to create a model of their new offices showing how efficient natural lighting can be in an office setting.

The Building

- Weatherize your building against heat loss in winter and air conditioning loss in summer. Patch cracks around doors and windows. Install additional attic, wall or subsurface insulation.

- Maintain your automatic doors so they open and close properly.
- Install awnings or plant trees for natural shade to block the sun's heat during summer.
- Install flexible windbreaks at your service entrances.

Heating, Ventilation and Air Conditioning

- Bring in a professional heating, ventilation and air conditioning specialist to be sure your system is working properly. New rooms or room dividers and more employees may be in your space since the system was installed or last repaired.
- Be sure the system is clean and dirt free, that all vents are properly set. You may want to turn the system down at off hours.
- If your system is old, you may want to invest in a new, more efficient system.

Equipment and Machines

- Buy and install modern high-efficiency electric motors when replacing older ones.
- Be sure your compressors have no leaks and you turn them off when not in use.
- Set your cooking equipment only to the highest temperature needed.
- Set your refrigeration equipment only as low as necessary. Install flexible clear plastic covers on cold display cases and be sure to completely cover them and open freezers when the store is not open to customers.

Energy Consumption and Air Pollution

Most of the electricity generated in the United States is created by burning coal, fuel oil and natural gas to drive turbines that produce electricity. Half of the energy used to make electricity is lost as heat in the process. Utilities and private generators are working to develop more efficient turbines and cogeneration systems that use the heat as well as the electricity. But until these developments come on line, the best way to maximize energy efficiency is to use the power wisely when it gets to you, the business consumer.

FUEL SOURCES

There are seven main fuel sources we can use to generate electricity: coal, fuel oil, natural gas, nuclear, hydroelectric, geothermal and solar. They have either environmental, health, technological or market-driven problems. Let's take a look at each one.

Coal

America's most abundant and native source of power. Almost one-third of the world's available coal is in the United States. That's 300 billion tons, enough to last another three hundred years. Coal is inexpensive. It reduces our dependence on foreign oil, which improves our balance of payments and strengthens our national security. Coal is used to produce half of our electricity needs and that is expected to increase. But coal is dirty, especially the hard coal mined in the Midwest and West Virginia. It is very high in sulfur, which emits sulfur dioxide into the atmosphere when burned. Sulfur dioxide is a prime contributor to acid rain.

A new report says that acid rain may not be as damaging to our forests as previously believed, but it is still a serious threat to water quality, fish, wildlife and productive agricultural land. It is also being linked to the deterioration and erosion of historic buildings and monuments in the Northeast.

Sulfur dioxide in the atmosphere is transformed into acid-sulfate aerosols, which have an adverse health effect on individuals. The American Lung Association suggests that sulfur dioxides contribute to excess mortality, increased risk of chronic respiratory disease and lung cancer. Respiratory diseases caused by air pollution are always most severe with older people and children.

One of the solutions to this problem is to burn low-sulfur coal mined in the western states. This would increase transportation costs and put a lot of miners in West Virginia out of work. Another solution is to install scrubbers or other new clean coal-burning technologies that remove most of the sulfur dioxides from the smokestack. These are expensive but effective.

But even if sulfur dioxide emissions are reduced, coal burning still contributes to the greenhouse effect and global warming by emitting carbon dioxide. The coal industry and many scientists believe the global warming theories are incorrect and overstated. But Michael Oppenheimer, a scientist with the Environmental Defense Fund, a leading authority on global warming, says that coal consumption should be cut in half to prevent a building up of greenhouse gases.

Fuel Oil

But coal isn't the only culprit in sulfur dioxide and carbon dioxide emission. Fuel oil contributes also—but not to such a great extent. There is also low-sulfur fuel oil that reduces sulfur dioxide emissions, although it is more expensive to use.

Fuel oil is also an abundant resource. The oil industry projects ample crude oil reserves already discovered will supply us up to the year 2030. But the United States is importing over 50 percent of its crude oil from overseas, mainly because it is cheaper to import oil than to pump it from American wells. Oil imports worsen our balance of trade, threaten our security and too often result in supertanker spills.

Furthermore, fuel oil delivery systems are already in place in the heavily populated Northeast. To change to coal or natural gas would cause dislocation problems.

Natural Gas

Natural gas is the cleanest burning of all the fossil fuels. It gives off virtually no sulfur dioxide emissions and significantly less carbon monoxide, nitrogen oxide and carbon dioxide than either fuel oil or coal.

The United States and particularly its neighbor Mexico have abundant supplies of natural gas. Unfortunately, most of the additional reserves are in the Middle East. Getting natural gas through pipelines from Texas is much easier than getting compressed natural gas on ships from foreign countries.

Furthermore, it is easier to build underground natural gas pipelines through less populated areas of Texas and Oklahoma than through densely populated New York or Connecticut.

Still, most environmentalists seenatural gas, because it creates so much less air pollution, as the fuel to rely on until solar power can become competitive.

Nuclear

Nuclear reactors produce from 3 to 5 percent of the electricity used in the United States. They don't give off any air pollution when they are in use, and for that reason they are getting a second look after their demise in the 1980s.

The nuclear industry has been plagued with construction problems, cost overruns, delays, safety and security regulations and environmental regulations that nearly put them out of business. No new nuclear plants have been applied for at the Nuclear Regulatory Commission since the late 1970s.

After the Three Mile Island and Chernobyl breakdowns, the public lost confidence in the industry. Permanent storage and containment sites for spent nuclear fuel cannot be found or built for both political and scientific reasons.

The National Research Council, an agency of the National Academy of Sciences, says that an underground nuclear waste dump that will be safe for ten thousand years, according to federal policy, is impossible to build.

Congress has designated a site in the Yucca Mountains of Nevada as a study site. It has yet to be certified by the Nuclear Regulatory Commission or the Environmental Protection Agency. At present, most high-level nuclear waste—that is, spent fuel—generated by power plants is being stored on-site at the plants in either water pools or dry storage. Low-level waste such as filters, rags, maintenance worker clothing, is being disposed of at three sites in South Carolina, the state of Washington and Nevada.

The French have a much greater dependence on nuclear-powered electricity than we have, in excess of 25 percent of their total power demand. They, too, have been storing the waste on-site but are now in the process of vitrifying their waste—a process akin to turning the waste to glass.

Hydroelectric

Water power has been used in the United States since the Dutch began building mills on streams in the Hudson Valley in the early 1600s. The old mill stream holds a warm place in the hearts of many Americans. Modern hydroelectric power is one of the reasons that New York City electricity rates are relatively low.

Hydro is a valuable but limited source of electricity and it does cause some environmental degradation. During the energy crisis of the 1970s, "low head hydro," meaning that electricity could be generated from slower-moving streams such as the Mississippi River rather than exclusively from faster-moving rivers such as the St. Lawrence, became a possibility because of increased demand for alternative power sources.

Geothermal

"Geothermal" power means using the heat from under the earth as a source of heat for electrical power plants and for direct use in heating and cooling. Most geothermal used for electricity is centered in the western third of the United States, California,

Nevada, Idaho and Utah, where geothermal temperatures are high enough to be used to generate electricity.

These plants produce 2,700 to 3,000 megawatts of power, equal to three to four nuclear power plants. Geothermal electricity has more than doubled in the 1980s because federal regulations have made it competitive for entrepreneurs to develop these power sources and sell the electricity to larger utilities such as Pacific Gas & Electric and California Edison.

Geothermal produces practically no by-products affecting the ground environment, the atmosphere or the ozone layer.

In addition, 17 trillion Btu's of direct use geothermal are used to heat greenhouses, hospitals, fish farms and other buildings in many areas of the West. And, finally, geothermal heat pumps are becoming an important heating and cooling competitor in the Midwest, where as many as twenty thousand devices are being installed each year. These heat pumps work like air-drawn heat pumps but instead draw their heating and cooling temperatures from the ground beneath the house.

These geothermal heat pumps are so efficient and energy-conserving that Public Service of Indiana is offering incentives up to $2,000 per unit for developers to install them because building new power plants for conventional heating and cooling is twice as expensive.

Solar

Solar power has taken great technological strides since its heyday in the mid-1970s. Then, loaded with high energy costs and federal income tax credits, solar started making inroads in the residential market with hot water heaters and hot air collectors.

But there were too many fly-by-night solar contractors doing shoddy work, and consumers lost confidence. That, coupled with a decline in energy costs, almost caused solar's demise in the 1980s. Many creative minds and entrepreneurs left the field.

Now, with new environmental concerns and the difficulty utilities have building new generating capacity, solar has started to make a comeback.

Out in the middle of the Mojave Desert, southeast of Los Angeles, the Luz Corporation has built eight solar thermal collectors and is selling the electricity to Southern California Edison. The collectors are huge curved mirrors that track the sun, collecting its energy. The heat from the sun is used to turn a turbine, which generates enough electricity to serve 300,000 residential consumers.

Southern California Edison has signed a thirty-year contract to buy the electricity. The problem is that solar costs 8 cents per kilowatt hour to generate, compared with 6 cents per kilowatt hour for coal or oil-fired electricity in the Southern California market. Nationally, the average cost of fossil-fueled electricity is 8.3 cents an hour, but solar thermal is not always practical in those parts of the United States that don't have as much sunlight and open space as California.

But if states start restricting the amount of air pollution utilities can generate using fossil fuels and if oil prices start to rise as expected, solar can become even more competitive, giving them a chance to find more investors, build bigger plants and generate more electricity.

The Mojave Desert collectors are called "solar

thermal" because they still must generate heat to create electricity. Another type of solar power is "photovoltaic," which creates electricity chemically when sunlight hits specially treated solar cells.

Photovoltaic solar, or pv, was developed in the 1950s and has been used in the space program. As pv technology has improved, the cost of generating electricity has dropped from 90 cents per kilowatt hour to 25 to 30 cents per kilowatt hour. Manufacturers of pv cells are trying to modernize their production techniques to mass-produce the cells, making their cost more competitive. The Solar Energy Research Institute in Golden, Colorado, expects the price of pv energy to drop to 5 cents per kilowatt hour by the year 2030.

Photovoltaic solar electricity will have its greatest application and market penetration in generating peak power. On extremely hot and sunny summer days, utilities have to start up their reserve generators to meet the heavy peak demand when everybody turns on their air conditioners. These sunny days are ideal conditions to produce short-term electricity from photovoltaic solar.

Another area in which pv cells can become popular is on electric automobiles. The problem with electric automobiles is that they have to be recharged after only 150 miles. On sunny days, pv cells on cars can generate electricity directly from the sun, keeping the car moving without so many recharges.

POLLUTION PREVENTION AND THE CHEMICAL INDUSTRY

One of the major new trends in corporate environmentalism has been the emergence of pollution prevention. Since the Clean Air and Clean Water Acts were enacted, companies have been focusing on pollution control. In other words, after the pollution is created in their manufacturing process, the company spends money to keep it from polluting the air, earth and water.

Pollution prevention is the industrial equivalent of residential source reduction. Companies and consumers change their habits and methods, cutting down on the amount of waste that they create in the first place.

According to the White House Council on Environmental Quality, "In the past several years, economic incentives for firms to prevent the generation of wastes have strengthened significantly for several reasons. First, pollution control, cleanup, and liability costs are increasing. Second, costs of resources inputs—energy and raw materials—are increasing also, further encouraging their efficient use. Finally, public pressure has increased, magnified by new requirements under section 313 of the Superfund Amendments that require firms to report annual releases of toxic chemicals."

The environmental quality report lists five pollution prevention techniques and examples:

1. *In-process recycling.* Reusing materials on-site that previously had been discharged. 3M saved $500,000 and reduced disposal of solvents by recycling 95 percent.
2. *Process modification.* Minor changes in technology or equipment can produce benefits. Exxon changed the roofs on some chemical storage tanks at a site in New Jersey and reduced their emissions by 680,000 pounds in 1983 alone.
3. *Improved plant operations.* Better maintenance and handling procedures, environmental audits and employee training can prevent pollution. Chevron

USA was able to reduce by 90 percent its chromium and nickel discharges into San Francisco Bay after an environmental audit pinpointed their source.

4. *Input substitutions.* Using less hazardous raw materials in a product can reduce pollution. If the coatings industry could shift from chemical to water-based solvents, thousands of tons of volatile organic compounds could be kept out of the air.
5. *Changes in end products.* Redesigning products for use at the consumer level. Putting hair spray in a pump applicator rather than an aerosol is an example; so is using unbleached paper.

Several large chemical companies have been very innovative in their approach to pollution prevention and environmental cleanup. Du Pont, 3M, Monsanto and Dow have been leaders in environmental awareness among large corporations.

3M, St. Paul, Minnesota: Pollution Prevention Pays and 3P+

3M's 3P program, which means Pollution Prevention Pays, has been widely heralded as a model in the chemical industry of how to prevent pollution before it is created. The program was started in 1975, and since then the company has saved $500 million by cutting the costs of pollution handling and disposal. They have also prevented the release of 112,000 tons of air pollutants, 15,300 gallons of water pollutants, 1 billion gallons of wastewater, and 397,000 tons of sludge and solid waste.

3M's 3P program encourages technical employees who have good ideas about how to prevent pollution to inform top management about their strategies.

They encourage employees to present ideas for ways to redesign equipment and products to cut down on pollution. Since the program started, they have initiated 2,511 3P programs worldwide. Some examples:

- 3M's Office System Division in White City, Oregon, redesigned a dry oven used in the manufacture of chemically coated paper for the microfilm users. They invested $16,000 and saved $533,200 annually, and prevented 137 tons of solid waste and 53 tons of air pollution.
- 3M's Riker Laboratories in Northridge, California, developed a water-based coating to replace a solvent coating for their medicine tablets. They invested $60,000 in the process but saved the need to buy $180,000 worth of pollution control equipment. They prevented twenty-four tons of air pollution annually.

In 1990, 3M developed a continuation of 3P called 3P+, which commits the company to further reduce all 3M manufacturing emission by 90 percent by the year 2000. 3M has already started spending $150 million to add additional pollution control equipment at 3M facilities worldwide.

E. I. du Pont de Nemours and Company, Wilmington, Delaware

Du Pont is well known as a leading producer of ozone-depleting CFCs. When released into the air, CFCs rise to the upper atmosphere, where sunlight breaks them down, releasing their chlorine molecules, which deplete the earth's protective ozone blanket.

Freon is Du Pont's trade name for the ozone-depleting refrigeration fluid commonly in use. In April

1990, Du Pont started a Freon recycling program in which they will receive spent Freon and other CFCs from companies, and then purify it, if possible, and reuse it.

Furthermore, Du Pont announced in June 1990 that they would build four plants to produce refrigeration chemicals that do not endanger the earth's ozone layer. The new compounds, hydrofluorocarbons or HCFCs, do not destroy the ozone layer.

E. S. Woolard, chairman of Du Pont, has made environmentalism an important part of his corporate outlook. In 1989, he told members of the American Chamber of Commerce that "our continued existence as a leading manufacturer requires that we excel in environmental performance.

"Environmentalism is now a mode of operation for every sector of society—industry included. We in industry have to develop a stronger awareness of ourselves as environmentalists. I am personally aware that as Du Pont's chief executive, I'm also Du Pont's chief environmentalist."

Woolard told the World Resources Institute in 1989 that "corporations that think they can drag their heels indefinitely on genuine environmental problems should be advised: society won't tolerate it, and Du Pont and other companies will be there to supply your customers after you are gone." Woolard has called on his colleagues to develop an attitude of "environmental stewardship" and "corporate environmentalism."

Du Pont's environmental goals include:

- Thirty-five percent reduction at the source of total hazardous waste by 1990.
- Create wetlands and other habitat for wildlife on one thousand square miles of land under their control.

- Eliminate heavy metal pigments in some plastics.
- Build more plastics recycling facilities.

Dow Chemical, Midland, Michigan

In 1986, Dow formalized their waste reduction program and called it WRAP, for Waste Reduction Always Pays. WRAP is an in-house program that has decreased Dow's air emissions by 85 percent and losses to water by 95 percent. They invested nearly $6 million for forty-two WRAP projects in 1988 and 1989. They inventoried all losses to air, water and land, set waste reduction goals and priorities and tracked their progress after instituting changes.

In 1990, Dow started an office paper recycling program called PRO (Paper Recovery Operation) at their offices in Michigan. They expect to conserve 81,000 cubic feet of landfill space and save 17,000 trees.

Monsanto, St. Louis, Missouri

In June 1988, Monsanto Chemical Company in St. Louis announced that they would cut their emissions by 90 percent by the end of 1992. By the end of 1988, they had reduced their emissions by almost 17 percent.

Monsanto's chairman and chief executive, Richard J. Mahoney, has taken a public position supporting corporate environmentalism. In his remarks to the National Wildlife Federation in January 1990 he said: "The torch of environmentalism is being lit in corporate America. . . . Some have argued that sustainable development on a global basis is impossible. . . . Sustainable development may be as simple a concept as an expression you hear often in the Midwest: 'Don't eat your seed corn; don't eat next year's crop' . . . it means sustaining the resources to survive; air, water, land, minerals and animal and plant life."

To this end, Monsanto has developed what is being called the Monsanto Pledge:

- To reduce all toxic and hazardous releases and emissions, working toward an ultimate goal of zero effect.
- To ensure no Monsanto operation poses any undue risk to our employees and our communities.
- To work to achieve sustainable agriculture through new technology and practices.
- To ensure groundwater safety.
- To keep our plants open to our communities and involve the community in plant operations.
- To manage all corporate real estate, including plant sites, to benefit nature.
- To search worldwide for technology to reduce and eliminate waste from our operations, with the top priority being not making it in the first place.

THE CHEMICAL INDUSTRY AND RESPONSIBLE CARE

Only the chemical companies have taken more heat for pollution problems than oil companies. According to the Chemical Manufacturers Association newsletter, only 28 percent of the population looks favorably on chemical companies. That's below organized labor and nuclear power, and only slightly higher than tobacco companies. "Twenty years ago, business, for the most part, ignored Earth Day, figuring if you just ignored it, it would go away and not bother you," said Jon Holtzman, vice president for communications for the Chemical Manufacturers Association. "What we found was that while Earth Day itself was not the cause of environmental laws that came in the 1970s, it was a clear signal that the public was aroused and wanted something done," added Holtzman.

To respond to the public, the CMA has developed a program called Responsible Care that is designed to improve their image and performance on environmental issues. The guiding principles for Responsible Care are:

1. To recognize and respond to community concerns about chemicals and our operations.
2. To develop and produce chemicals that can be manufactured, transported, used and disposed of safely.
3. To make health, safety and environmental considerations a priority in our planning for all existing new products and processes.
4. To report promptly to officials, employees, customers and the public information on chemical-related health or environmental hazards and to recommend protective measures.
5. To counsel customers on the safe use, transportation and disposal of chemical products.
6. To operate our plants and facilities in a manner that protects the environment and the health and safety of our employees and the public.
7. To extend knowledge by conducting or supporting research on the health, safety and environmental effects of our products, processes and waste materials.
8. To work with others to resolve problems created by past handling and disposal of hazardous substances.
9. To participate with government and others in creating responsible laws, regulations and stan-

dards to safeguard the community, workplace and environment.

10. To promote the principles and practices of Responsible Care by sharing experiences and offering assistance to others who produce, handle, use, transport or dispose of chemicals.

Any company that wants to be a member of the Chemical Manufacturers Association must take the pledge of Responsible Care.

For more information, contact the Chemical Manufacturers Association, 2501 M Street, NW, Washington, D.C. 20037, 202-887-1260.

In addition to becoming more responsible about the impact their products can have on the environment, chemical companies have started to reach out and meet environmental groups on common ground. The environmental groups have responded and have begun recognizing the efforts of the chemical industry. The Corporate Conservation Council and its awards are an example of that change.

CORPORATE CONSERVATION COUNCIL—NATIONAL WILDLIFE FEDERATION

The Corporate Conservation Council was established in 1982 and gave its first awards in 1985. The purpose is to bring industrial leaders into contact with environmentalists to reduce or avoid conflicts by "exchanging information and examining proposals for conserving environmental values in concert with sustainable economic development."

Board members of the council include ARCO, Browning-Ferris Industries, CIBA-GEIGY Corporation, Combustion Engineering, Dow Chemical, Duke Power Company, E. I. du Pont de Nemours & Company, General Motors, 3M, Monsanto, Shell Oil, USX, Waste Management and Weyerhaeuser.

In 1990, the Environmental Achievement Award was presented to 3M for its innovative waste reduction program, Pollution Prevention Pays. They had prevented the release of 538,000 tons of pollution annually since 1975.

In 1989, the award went to Tennessee Chemical Company for its soil conservation efforts at the Copper Basin Reclamation Project in Polk County, Tennessee. They reclaimed 12,000 acres of environmentally degraded land mined in the 1850s.

In 1988, the award went to International Minerals & Chemical Corporation's Peace River Floodplain Habitat Wetlands Preservation Project in Bartow, Florida. They are transforming 525 acres of strip-mined land into a wetland wildlife habitat.

In 1987, the award went to Amoco Corporation for their North 40 wetland water purification and wildlife sanctuary at their oil refinery in Mandan, North Dakota.

In 1986, the award went to Agrico Chemical Company for creating wetlands at the company's Florida and North Carolina sites.

To nominate a program for an award, contact the National Wildlife Federation's Corporate Conservation Council at 1400 Sixteenth Street, NW, Washington, D.C. 20036-2266.

THE BOTTOM LINE

No one is going to stand up and say the chemical industry has not been environmentally neglectful over the history of its existence. There have been abuses, there have been accidents and there has been pollution.

But it is clear today that because of enlightened corporate leadership, a willingness to see that the bottom line and the environment have a direct relationship with each other, and in many cases because government and the public have held their feet to the fire, the chemical industry is now committed to a cleaner environment.

5

Transportation

The main cause of air pollution in New York is vehicular.

—Bill Hewitt, New York State Department of Environmental Conservation, *New York* magazine, April 16, 1990

People in the United States and around the world were appalled when the *Exxon Valdez* ran aground and spilled nearly 11 million gallons of crude oil into the pristine waters off the coast of Alaska.

The fire on board the Norwegian tanker *Mega Borg* near Galveston, Texas, and the numerous oil spills in the New York harbor in the spring of 1990 have created anger and confusion in people who depend on petroleum products to power their vehicles.

People are ready, willing and able to criticize the oil industry for all the environmental degradation associated with exploration, drilling, production, transportation and distribution of petroleum fuels. But the environmental organization Greenpeace points the finger in a different direction.

They have designed an ad that looks like a wanted poster with a black-and-white photograph of the captain of the *Exxon Valdez* in the middle of the page. The headline reads: "It Wasn't His Driving That Caused the Alaskan Oil Spill. It Was Yours."

And that's the point. As long as we are "a nation drunk on oil," according to Greenpeace, we are going to have oil spills and we are going to have air pollution.

The energy crisis of the 1970s showed us that driving fuel-efficient cars using cleaner-burning fuels, car pooling and taking public transportation does not mean we are going to lower our standard of living, productivity or profit margin.

We have a new energy crisis in the 1990s and it is called the environment. During the 1970s air quality improved. But now all of the gains we made during the 1970s have been eclipsed because there are too many people driving too many cars. While we wait for the automakers and the oil producers to bring us smog-free driving, there are a number of things we can do to clean up air pollution right at the place where we work.

There is no denying that people moving back and forth to their place of work is a primary source of air pollution in the United States. Thirty percent of all vehicle miles traveled in the United States are for the purpose of commuting. Business can take responsibility for the air pollution created by commuting and dramatically reduce it. Business can get out front of this and take the lead. Traffic congestion is already a productivity sink. It is stressful and time consuming for people to commute by car.

In this chapter, we are going to take a look at just how bad air pollution has become in our metropolitan areas. We are going to analyze smog and detail the health effects on our people. Vehicular air pollution is also a primary cause of global warming and the greenhouse effect that needs to be prevented.

The first solution we will look at is efficiency, how a corporate transportation awareness program can make your fleet more efficient, reduce its emission of CFCs and VOGs, cut down on discarded batteries, tires and motor oil, as well as getting your employees to maintain their own cars the way you maintain the company cars.

The second solution is a look at the new fuels and technologies now being introduced or proposed for common use: reformulated gasoline, methane, ethanol, natural gas, plus electric cars, how they are being put to work now and how they will work in the future.

The third and most important solution is simply to get people out of their cars. We will look at efforts in public transportation, as well as techniques and strategies such as van and car pooling, rails to trails, bicycling, walking and other initiatives to reduce vehicle miles traveled, and therefore the amount of air pollution your business creates.

Look at it this way. When employees commute back and forth to work each day, you and your business create and control a certain amount of air pollution. Air pollution degrades the environment, worsens the health of you and your employees and their families. Traffic congestion costs you and your employees money. It creates stress and aggravation, which can lower productivity.

Your response to this crisis can make your company a desirable place to work, increase productivity and profits, help you in recruiting the best employees and make your city or region more desirable for new business start-up or relocation. Business has a tremendous environmental and economic stake in transportation efficiency and clean air.

TRANSPORTATION'S EFFECT ON THE ENVIRONMENT AND HEALTH

Photochemical oxidant, better known as smog, was first identified in Los Angeles in 1943. The peculiarities of Los Angeles, a geography that encourages trapped and stagnant air, plenty of sunlight and lots of cars, gave smog a chance to get started.

Smog is composed primarily of ozone, which is caused by the chemical interaction of tailpipe emissions and heat from sunlight. The two tailpipe emissions are hydrocarbons and nitrogen oxides, both released as by-products of the gasoline-fueled internal combustion engine. Gasoline itself evaporates and gives off hydrocarbon vapors, contributing to smog.

Automakers have helped reduce smog over the years by inventing and installing catalytic converters and building cars that are lighter and more fuel efficient. But since the end of the energy crisis and the stabilization of gasoline prices in the 1980s, American motorists have started driving more miles, driving more cars and driving bigger, heavier cars that get fewer miles per gallon. The result: all the clean air advances made during the 1970s are about to be overtaken by a new wave of smog, threatening our health and our environment.

Health Effects—The health effects of smog are more acute in people who are susceptible to respiratory problems in the first place. Older people and younger children are most at risk. The American Lung Association reports that exposure to smog can cause eye, nose and throat irritation, coughing, shortness of breath, headaches, tightening in the chest, impaired pulmonary function, altered red blood cells and can trigger asthma attacks for those afflicted with that condition.

Carbon Monoxide—CO is a colorless, odorless toxic gas that also comes from tailpipe emissions. It limits the body's ability to transport oxygen to body tissues and places a strain on people with heart conditions, elderly people and pregnant women and their unborn children.

The Greenhouse Effect—Another product of tailpipe emissions is carbon dioxide, CO_2. Up until now, CO_2 has been considered a relatively harmless gas. In fact, when you breathe, the air you exhale contains CO_2. It is a natural by-product of the combustion process. In the symbiotic ecosphere, CO_2 is fuel for green plants and trees that take in CO_2 and give off oxygen. Sounds pretty harmless.

The CO_2 problem we are facing now is the discovery of the greenhouse effect and its resultant global warming. Carbon dioxide gas given off during the burning of coal, wood, fuel oil and gasoline rises in the atmosphere and forms a gaseous trap. It allows heat from the sun to penetrate to the earth's surface, but when this heat from the earth rises and tries to dissipate, the greenhouse of gas stops it. The result is a gradual warming of the earth's surface.

There is an enormous difference of opinion about the validity of a global warming trend, but those who support the theory predict that the earth's temperature will rise, causing climatic zone changes, more droughts and deserts, resulting in a diminished capacity of agriculture, a melting of polar ice caps and a gradual rising of sea level, creating havoc for major port cities like London, New York and Hong Kong.

There are other causes of increased carbon dioxide levels in the atmosphere: the burning of the South American rain forests and the increased urbanization of America, both of which mean there are fewer green plants and trees to consume the CO_2. We certainly need to conserve more open spaces and plant more trees, but we also need to produce less CO_2 and other tailpipe emissions if we are going to have any impact on cleaning up our dirty air.

SOLUTION TO POLLUTION NUMBER 1: Increased Efficiency, or How We Can Do the Most with What We Already Have

It is one thing to talk about increased vehicle efficiency, but another to make it happen within your company fleet of cars and/or the individually owned fleet of cars driven by your employees as commuting vehicles. You have to have a program that will accomplish your goals.

1. First, appoint a high-ranking company officer or executive to be responsible for the creation, implementation and monitoring of your program. If your company is big enough, you might want to hire a full-time staff person, or bring in a consultant. In either case, you must have an executive with authority and access overseeing the project.
2. Analyze your corporate- or employee-owned fleet. How many cars are involved? How many miles are

driven each year that are commuting related? How many miles are driven that are job related? Break that down into employee vehicle miles traveled. This gives you a bench mark number that can be measured and improved upon.

3. Analyze the types of cars in the fleet by fuel efficiency and quality of miles driven; in other words, are most of your fleet miles driven by commuters in congested low-efficiency traffic, or do you have a lot of long-haul salespeople or staff making calls by car? This gives you another number that can be improved on.
4. Set corporate goals for fuel efficiency in automobiles. New York State just raised their fuel-efficiency requirements for state-owned automobiles to 29.9 miles per gallon from 27.5. The Environmental Protection Agency has a *Gas Mileage Guide* booklet that can help you know which cars get which mileage. For instance, some of the prestige cars like the BMW 535i, Volvo 760, Porsche 911 or Chrysler Fifth Avenue get less than seventeen miles per gallon in city driving, while the Honda Civic CRX, admittedly a much smaller car, can get fifty miles per gallon in city driving.
5. Buy the most fuel-efficient cars you can that still meet your needs. Make your reasons for buying known to your employees and encourage them also to buy a fuel-efficient car and tell them why.
6. Start a program of annual vehicle inspections within your company if your state doesn't already have one. Most state inspections are required for safety reasons and not pollution control and fuel efficiency. The American Lung Association reports that vehicles that are inspected annually can reduce their emission by 25 to 30 percent, and reduce their fuel use by twenty gallons each year.

A vehicle inspection would include testing tailpipe emissions, but also check for efficiency in spark plugs, fuel injectors or carburetors, air filters, tire wear and auto air conditioning, a leading cause of CFC leakage, resulting in stratospheric ozone depletion, which some suspect is a leading cause of skin cancer and other disorders.

THE SOLUTION TO POLLUTION NUMBER 2: New Fuels

Fuel efficiency is an important and productive first step in reducing your business's contribution to air pollution. But many people feel that gasoline needs to be changed and substitutes for gasoline need to be put to use to take another major bite out of air pollution.

People are now looking at a whole range of new fuels, including reformulated gasoline, ethanol, methanol, natural gas, and even electricity, that will power our vehicles but produce less air pollution. Each has its strengths and weaknesses and each is currently in use or ready to come off the drawing board.

Reformulated Gasoline—This is the fuel that the *New York Times* called "the health-food equivalent of gasoline." Others have dubbed it "green gasoline." As we mentioned before, the problem with burning gasoline is that it gives off carbon monoxide gas and unburned hydrocarbons, which contribute to ozone, the prime component of smog.

Reformulated gasoline is a gasoline-based fuel whose formula has been adjusted to reduce pollution from motor vehicles. Reformulated gasoline contains substantially less aromatics and olefins, two of the prime contributors to atmospheric ozone.

Also, reformulated gasoline contains less butane, which helps retard the evaporation of gasoline before it is burned, and less benzene, a toxic aromatic, which is another hydrocarbon that is difficult to burn completely.

Reformulated gasoline will require petroleum refineries to shift their operations to produce more alkylate gasoline, a gasoline that contains no aromatics, including benzene, no olefins and very little butane.

STRATCO, Inc., of Leawood, Kansas, the world's leading licensor of the alkylation process, estimates that increasing the alkylate content of gasoline from the current 12 percent to 30 percent will allow all gasolines to be reformulated gasoline.

Finally, reformulated gasoline contains methyl tertiary butyl ether or other oxygen-containing additives. These oxygen-rich additives make the fuel burn more cleanly, giving off less carbon monoxide and hydrocarbons.

ARCO, the Southern California oil giant, introduced a reformulated gas called EC-1 in September 1989. EC-1 can be used only in pre-1975 cars that do not have catalytic converters and burn regular leaded gas. ARCO continues to work on developing a reformulated gas for use in all cars, but EC-1 is credited with having reduced emission from pre-1975 cars in Southern California by 20 percent.

Shell Oil Company has introduced SU 2000E as a premium unleaded gas in Los Angeles, San Diego, Chicago, Milwaukee, New York, Hartford, Philadelphia, Washington and Baltimore. The Clean Air Act has targeted all of those cities except Washington for the introduction of cleaner-burning fuels.

Marathon Oil Company introduced Amaraclean gasoline in the Detroit area in March of 1990. Amaraclean gasoline comes in all three unleaded categories, regular, mid-grade and premium.

Phillips and Conoco have been test marketing reformulated gas in Denver, St. Louis and other areas.

The main problem with reformulated gasoline right now is the cost to produce the additive methyl tertiary butyl ether, MTBE. MTBE has been used as a replacement for lead in gasoline to boost octane for over twenty years. But the new reformulated gasolines demand a much higher ratio of MTBE, creating greater demand for the product. Oil companies say it will cost them billions of dollars to build new refining capacity to meet the demand for MTBE. They are reluctant to make the investment until they see how new state and federal clean air laws are drawn and whether consumer demand is high enough.

Ethanol—Ethanol is the new fuel favored by the midwestern Corn Belt states because millions of bushels of corn will be used to produce the grain alcohol that is blended with gasoline to make it. It is better known as gasohol in the Midwest, where some service stations selling ethanol proudly display a corn emblem on the pumps. Half the fuel used to power cars in Brazil is ethanol.

Ethanol can be used in existing auto engines without any modifications, and it emits less carbon monoxide and less carbon dioxide than gasoline. It does emit aldehydes, which can contribute to smog.

The chief disadvantages of ethanol is that it is energy intensive and therefore expensive to produce. Ethanol became competitive during the oil crisis of the 1970s because of the high cost of imported oil. But when the cost of oil dropped, so did ethanol's competitive edge.

Furthermore, greater demand for ethanol could mean increased corn production, which demands intensive fertilizer, herbicide and pesticide use, a prime source of water pollution. To be truly competitive, ethanol depends on continued government subsidies. These have been reduced significantly in the last few years and gasohol has been strugging to maintain a position in the marketplace.

Methanol—Methanol, or wood alcohol, is produced primarily from natural gas in the United States. It can also be produced from coal and wood from forest residues. Ethanol burns cleaner than gasoline, giving off fewer emissions that contribute to smog.

Methanol does give off significant emissions of formaldehyde, which is thought to cause cancer and other health problems. If methanol is made from coal, there would be an increase in the emission of carbon dioxide, a greenhouse gas. Furthermore, existing car engines cannot run on both gasoline and methanol at the same time, so modifications would be costly. Methanol gets about half the mileage of gasoline, so twice as much would have to be used.

Natural Gas—Compressed natural gas is a very clean-burning fuel that gives off little in the way of hydrocarbons or carbon monoxide. It still does give off carbon dioxide and nitrogen oxide.

One of the benefits of using CNG is that with a slight modification, existing engines can use either CNG or gasoline.

■ The Brooklyn Union Gas Company in New York has

over 200 of their company vehicles fitted to use either gas or CNG.

- The United Parcel Service is going to convert its 2,700 delivery trucks in the Los Angeles area to run on natural gas over the next few years.
- General Motors and a group of natural gas companies in California, Colorado and Texas are designing and building a fleet of trucks to be powered by natural gas in 1991.

One of the drawbacks of CNG is that it is quite heavy and requires the installation of larger high-pressure storage tanks to carry the load.

Electricity—Battery-powered vehicles fueled with electricity in and of themselves produce no pollution. If all vehicles in the United States were electrified there would be significant reduction in both air pollution and greenhouse gases. The United States would end its dependence on foreign oil and improve its balance of trade deficit. Scientists say we have sufficient electricity capacity to generate the electricity need to power electric cars. The air pollution problem has become so bad in some areas, such as Southern California, that General Motors, Ford and Chrysler are all working to develop electric cars.

The disadvantages of electric cars are their range, only about 120 miles before they need to be recharged, the huge number of batteries that will be needed over the life of the car and the potential for expanded air pollution from coal-fired power plants. GM's prototype electric car Impact has thirty-two lead acid batteries that must be replaced every twenty thousand miles. All those toxic batteries in the waste stream will cause landfill nightmares until battery recycling becomes a reality.

If cleaner coal or cleaner coal-burning technologies can be introduced, as well as a solution to the radioactive waste generated at nuclear power plants, and if solar cells and more hydroelectric power can be harnessed, electric cars will become very desirable to consumers and competitive in the market.

THE BOTTOM LINE: The bottom line on new fuels is that reformulated gasoline and ethanol are the two top contenders. The Clean Air Act will call for some sort of cleaner-burning gasoline to be in the pumps in the smoggiest cities in the United States. Oil companies want reformulated gas and the corn states want ethanol.

Electric cars have a strong potential in the same areas where new fuels would be introduced—the smoggiest urban areas. Commuters, delivery vehicles and "suburban run-around" drivers are likely to have the most incentive to buy electric cars.

SOLUTION TO POLLUTION NUMBER 3: We've Got to Get Out of Our Cars

Increased efficiency and cleaner-burning fuels all have dramatic potential but ultimately limited ability to reduce air pollution and clean up the environment. Cheap fuel, cheap cars and an affluent and expanding population mean that improvements in automobiles and fuels can never keep up with the number of people who are buying cars and using them.

Only 30 percent of vehicle miles traveled are for the purposes of commuting back and forth to work.

But that is a significant number, and it is the one number that business can do something about. Business can and should be committed to reducing the amount of pollution it is responsible for.

The way to do that is to get people out of their cars. And there are a great number of ways to accomplish this, including car and van pooling, public transportation, busing, even walking and bicycling in certain situations.

First, designate a high-level executive of the company to oversee the project. They don't have to do all the work, just make sure somebody with clout has responsibility.

Next, analyze your workforce's commuting habits, what methods they use to get back and forth to work.

Third, analyze the demographics of your workforce, where they live in relation to major arterial roads and public transportation systems.

Finally, set goals for reducing the amount of air pollution your employees create in commuting. Getting people to change their driving habits is extremely difficult, but if you pose it as a way to protect the environment, you have a new tool that should get people to respond. Your chances for success will be greater if you decide to use a mix of transportation alternatives rather than relying on just one.

Car Pooling and Ride Sharing or "Clean Air Company Cars"

The easiest thing you can do to reduce employee vehicle miles is to start a car pool at work. A computer analysis can match up people who live near each other or on direct routes to the workplace. Employees who car pool can be given priority parking spots nearer to the front door.

Van Pooling or "Clean Air Shuttles"

Instead of having to drive their own cars in a car pool, employees could be organized into van pools. Existing models for van or bus pools are school buses and express buses that run at rush hours as well as "airporter"-type vans that ferry airline passengers from the airport to their hotels or homes.

Employees could pay for their rides or the company could defray part of the cost. You could buy, rent or lease the vans or buses yourself, but you might want to arrange a deal with a local transit company to run your employee shuttles for you.

The city of Los Angeles offers a $5,000 per van subsidy to any employer with a work site located in the city who will start and run a van pool program for at least four years. Van or bus pools or employee shuttles could also have priority pickup and delivery at the front door of the building.

A van pool is the most efficient way to move employees because it only has to run at peak hours and can always be filled to capacity.

The Parking Problem or "The Environmental Parking Lot"

Free parking provided for employees amounts to a $400-a-month tax-free benefit for them. It can cost from $5,000 to $20,000, with $10,000 being the average, to build a single employee parking space at the workplace. It also costs from $75 to $300 a month to maintain the spot, with $150 being the average. Most zoning boards require that a certain number of parking spaces be built to handle employee cars at business sites.

Furthermore, the cost of parking spaces is tax deductible either to the business or the developer, whoever owns the property. The tax-free benefit an employer can give to employees to take public transit is only $15. In an era of increasing need to reduce air pollution, this tradition should be changed.

An option that you might try is to negotiate with your planning board to offer a transit subsidy to your employees, such as car pooling or bus passes, to be used to offset the number of parking spaces you are required to build and maintain.

Another parking option you might try is to put priority parking for van poolers and car poolers up front and put all other parking spaces for single-occupancy vehicles toward the back or away from the front door.

Public Transportation

According to the American Public Transit Association, in the next twenty years, traffic congestion will increase 400 percent on the nation's freeways, delay time will increase by 5.6 billion hours, excess fuel consumption will increase by 7.3 billion gallons a year and air quality will continue to deteriorate. Already, sixty-eight cities fail to meet federal ozone standards, and nearly 150 million Americans now live in areas with severely polluted air.

On the other hand, every commuter who takes public transportation instead of an automobile prevents the release of sixty-three pounds of carbon monoxide, nine pounds of hydrocarbons and nitrogen oxides and one pound of soot into the atmosphere.

First, support the building of public transportation networks. Railroads, subways, light rail, trolleys and buses, especially buses that are powered by cleaner fuels like methane or natural gas, are all better for the environment than a constantly increasing car and roadway system.

Second, work with your public transit agency to create bus routes that fit the needs of your workforce. Many bus routes are unsuccessful because they no longer run along routes where people live and need to get to work, especially in suburban areas. Too often bus routes run into a central downtown area, which may or may not be where new business has located. Metro-North in New York City is now finding an increasing ridership of people who live in the city but commute to Westchester, New Jersey or Fairfield to work. Public transportation needs to respond to this change, but they need your help.

Third, provide subsidies for your employees who do take public transit. In Southern California, companies with more than one hundred employees that offer free or subsidized parking for any employee must offer a $15-a-month public transit commuting subsidy to all employees, and actually pay it to employees who use public transit. For smaller companies with fewer than one hundred employees, the city of Los Angeles pays $5 per month for each $15 or more transit subsidy that the company pays voluntarily.

Soft Paths

Over one-half of the people in America live within six miles of where they work. If we were in the Netherlands, we would be walking or riding our bikes to work from that distance. But there is no reason why we can't do that here. In fact, in many cases now because of increasing traffic congestion for automobiles, buses and public transit, walking or bike riding might get you to and from work faster and more quietly.

The city and business community in Seattle has embarked on an Urban Trails Program that transforms old abandoned rail lines and other paths into urban trails suitable for walking, biking or jogging. The program now has thirty miles of trails and as many as ten thousand people a day use it to commute to work.

According to Peter Lagerway, a transportation planner for the city of Seattle, businesses are lining up to build offices and other operations along the urban trails because having office space along the trails is becoming a number-one amenity in the area.

Lagerway says that clean air and commuting alternatives are a part of an overall environmental awareness that Seattle has started using as an economic development tool—it's a good way to attract business and qualified employees to Seattle. He calls it "The Rainier Factor," named after the famous mountain that can easily be seen from Seattle on a clear day.

The Urban Trails Program offers a quality-of-life benefit and it creates a reason why people and businesses want to do business there. The people of Seattle and King County recently passed a $117 million bond issue, $30 million of which is for urban trails.

Contact your local city planning or economic development office or transportation department to see if any urban trails programs are in operation or are on the drawing board. If they do run near you or could run near you, offer your support for the project and alert your employees to its availability.

Telecommuting

Because of the onslaught of the information age and the availability of sophisticated computer, fax, modem and telephone systems, many of your employees do not have to be at your work site at all to perform certain functions. A lot of design engineering as well as information processing is done on computer and can be done from home.

Look at your employee functions and determine which can be done by computer or telephone at the employee's home. Design a system that lets you feed information back and forth and see which of your employees would like to avoid commuting in the first place. Telecommuting is becoming another attractive employee recruiting tool, similar to flex hours and job sharing, that can help you attract good employees, especially parents who want to spend more time with their children.

THE BOTTOM LINE

The bottom line in this chapter is that business controls 30 percent of the air pollution created by automobiles and trucks. People are looking for ways to contribute to environmental cleanup. Cutting down on air pollution from our commuting vehicles is a way to do it.

6

Green Marketing in the Environmental Era

The American consumer's reawakened interest in a cleaner environment has created new challenges and new opportunities for marketers aiming to sell at the retail level. Because of the types and amounts of packaging they use, and because of some of the products they sell, supermarkets and restaurants, as well as mass market retail stores, have become the battleground for environmental consumerism.

Recent surveys have shown that consumers are eager to buy products that are less polluting and they are willing to pay extra to buy them.

- In a September 1989 *Advertising Age* poll, one thousand people were asked, "How concerned are you about damage to the environment caused by the manufacturers of consumer goods and packaging?" On a scale of 1 to 5 with 5 being the most concerned, the average concern ranking was a 4.14.

- In that same *Ad Age* poll, 96 percent of women and 92 percent of men said they would "make a special effort to buy products from companies trying to protect the environment."
- Ninety-six percent of men and 94 percent of women said they would be willing to give up some convenience to buy those products.
- Ninety percent of women and 87 percent of men said they would be willing to "pay more for products or packaging made environmentally safer."
- A poll by the Michael Peters Group, a New York packaging consulting firm, learned that 89 percent of Americans are concerned about the impact on the environment of the products they purchase, that 78 percent would be willing to pay more for a product packaged with recycled or biodegradable materials, and that 53 percent of the survey said they had changed a purchasing decision in the last year because of the product's impact on the environment.
- Eighty-three percent of Americans have changed their shopping or living habits to protect the environment, according to a *Newsweek* poll.

The old adage of selling the sizzle, not the steak, or sell the package, not the product, has taken on a whole new dimension.

Furthermore, politicians have latched on to banning certain consumer products and packaging, polystyrene in particular, as a way to respond to public pressure and create a better profile for themselves.

But some politicians are responsibly attacking the packaging problem because it does contribute to nearly 30 to 35 percent of all garbage. In the Northeast, where landfill space is becoming a scarce commodity, the Coalition of Northeastern Governors, including Connecticut, Maine, Massachusetts, New

Hampshire, New Jersey, New York, Pennsylvania, Rhode Island and Vermont, is working to develop a set of "preferred packaging guidelines," as a way to reduce the amount of packaging trash that has to be disposed of.

All of this consumer and political pressure has caused packagers and marketers to introduce products labeled biodegradable, photodegradable, environmentally friendly, ozone friendly, safe for the environment and other claims that are now being called into question.

The rush to respond to consumers' demand for nonpolluting products has created a backlash.

- The Minnesota attorney general Hubert Humphrey III has said, "The selling of the environment may make the cholesterol craze look like a Sunday school picnic."
- Humphrey has been joined by the attorneys general from at least six other states in investigating environmental claims on packages.
- The Federal Trade Commission and the U.S. Senate are both moving toward establishing standards for environmental claims.
- The New York Public Interest Research Group has printed a booklet, *Plagued by Packaging,* that cites specific companies and products, such as McDonald's wrappings, Kodak's disposable camera and Procter & Gamble's disposable diapers, as bad for the environment. They list the companies' toll-free telephone numbers and urge people to call and complain about the packaging.
- The Washington Citizens for Recycling in Seattle released its first Packaging Awards and Booby Prizes from an Environmental Point of View. The story was given a half page in the *New York Times*

and blasted blister packaging, L'eggs panty hose, General Mills and other major companies.

- Environmental Action, the Natural Resources Defense Council and several other environmental organizations got together last year and destroyed the image of degradable plastics at a highly publicized news conference covered by the major networks.

ENVIRONMENTAL LABELING PROGRAMS

By now the consumer is thoroughly confused. And so is the products and packaging industry. How can business retain consumer confidence on the environment?

Enter the in-store labeling programs, lead by Wal-Mart, Kmart and some supermarkets.

Wal-Mart. In July 1989, William R. Fields, executive vice president for merchandising and sales, wrote a letter to his vendors that said in part: "Wal-Mart is beginning to assume our corporate responsibilities as they relate to our environment—the water we drink, the air we breathe and the land that supports us. . . . Our customers want to know what we are doing to environmentally improve the manufacturing process, the products themselves and the disposal process of products and packaging. . . . Wal-Mart is committed to helping clean up the world we live in."

With that, Wal-Mart set the products and packaging industry on its environmental ear. They started their in-house labeling program and challenged their suppliers to answer these six questions:

1. Are your products or packaging recyclable?
2. Are they made from recycled materials?

3. Are packaging systems refillable or reusable?
4. Do you have a product concentrated to reduce package volume and waste?
5. Is the manufacturing process now safer for our land, air and water?
6. Are your products now safer for the environment?

So far, Wal-Mart has labeled over one hundred products in their program. They continue to carry a full line of products, but now, at least, consumers have a choice.

Kmart. Kmart began its own program and publicly launched it with a full-page ad in *USA Today* on the Friday before Earth Day. Their program will also include in-store signage of environmentally safer products as well as promoting the already accepted symbol for packages made of recycled fibers.

Furthermore, Kmart is using recycled office paper in its stores, offices and distribution centers, recycling oil from their auto centers and encouraging people to bring them used auto batteries so that they can be disposed of safely.

OPERATION GREEN SEAL

But many environmental and consumer groups want to establish their own green seal programs. I hesitate to make this listing because some may be out of business by the time you read this, but it is still instructive that these efforts are or were under way.

Most of these programs were inspired by the Blue Angel environmental program that has been successful in West Germany.

- Green Seal, headed by environmentalist Dennis Hayes, will put its own green seal on light bulbs, laundry cleaners, house paint, toilet paper and facial tissue. Will charge companies fees on a sliding scale to conduct the research necessary to gain the seal.

- Green Cross, offered by Scientific Certification Systems, an Oakland, California, testing laboratory. Will also charge for research done to earn certification.
- Good Earthkeeping Pledge. Competition for the same name and program from the Good Housekeeping Institute, an affiliate of *Good Housekeeping* magazine and the Good Housekeeping Institute in Gainesville, Florida. Both will give awards to businesses that are environmentally responsible and then charge the companies licensing fees to use the seal on its packages and in its promotions.

"I generally applaud the initiative shown by the green seal people because they are trying to work from the bottom up rather than from the top down," says Melinda Sweet, director of environmental affairs for Lever Brothers. "How are we going to set scientifically reliable standards? In others words, who is going to watch the watchers?"

Environmentalists, state attorneys general and product packagers generally do not favor the green seal programs.

Opponents of the green seal programs fear that they will run into the same outcry the American Heart Association did when they tried to establish their heart emblem program for healthy food. Critics felt there would be a conflict of interest because they, too, would be charging to use the emblem.

Many people feel that the product and packaging companies themselves should shoulder the responsibility of making their products and packages better for the environment. They feel that the government, the EPA, the FTC or the FDA should set standards on definitions for terms such as "degradability," "recyclable" and the rest.

COMPANY ACTION

In the meantime, companies are moving ahead in new product and packaging content and designs. Procter & Gamble, Lever Brothers, Colgate-Palmolive, Bristol-Myers, Heinz, General Foods, Kodak, are all moving forward on the environmental question.

Lever Brothers

Eco-Care Pack packaging is Lever Brothers new concept in packaging to reduce its share of the solid waste stream. It is a "Bag-in-Box" design that puts a plastic bag in a cardboard box. Both Snuggle liquid fabric softener and Wisk liquid laundry detergent are being packaged in this design. The plastic bag uses 75 percent less plastic than conventional bags and the cardboard box is corrugated, one of the most recycled products on the market today. In addition, Wisk liquid detergent is made with no phosphates and the surfactants and enzymes are biodegradable.

Beyond packaging, Lever Brothers ceased using pigments and inks containing heavy metals for plastic and paperboard packaging, they are streamlining or "lightweighting" plastic bottles, they've added the plastic industry's plastic coding to bottles to facilitate sorting for recycling and they use recycled paperboard for boxes of laundry soap, bar soaps and automatic dishwasher soaps.

They are experimenting with recyclable shipping boxes and packing plastic, and they are recycling waste heat into electricity at their cogeneration plant in Los Angeles.

But one of the most important things they are dong is the creation of their Packaging Development Center near Baltimore. The center is headed by a vice president, Arnold Brown, and staffed by forty scientists and engineers.

"Making our packages better citizens in the landfill is one of our top priorities here," says Brown. "We are looking at all the elements of our packaging design as they relate to recycling, removing heavy metals from our inks and pigments, the availability of using recycled materials in our packages and other environmental concerns."

Procter & Gamble

By some estimates, the products and packages produced by Procter & Gamble contribute nearly 2 percent of all the trash that goes to the landfill. Some estimates say P&G's Luvs and Pampers disposable diapers take up almost 1 percent of our trash all by themselves. P&G disputes these figures, as well as the consequences of the trash they do contribute. Nonetheless, they have taken great strides in streamlining their packaging as well as other initiatives.

Geoff Place, vice president for research and development at P&G, says, "Environmental quality is a consumer need that's as real as any our brands address, and we must learn to manage it in much the same way we manage other consumer needs, such as Tide's cleaning power or Luvs's leakage protection.

"Environmental quality is our opportunity to offer a new, additional benefit to consumers and customers. We're also making the earth a better home for ourselves and future generations."

Here are some Procter & Gamble environmental initiatives:

- Spic and Span Pine bottles are now made of 100 percent recycled PET plastic, the same plastic used in making soft drink bottles.
- Downy, Tide, Cheer, Era and Dash bottles are now made of at least 25 percent HDPE plastic, the same plastic used in making juice and milk bottles.

  These two plastics recycling initiatives are expected to keep 80 million plastic containers out of landfills. P&G will use more recycled plastic when recyclers can meet their demand. This should spur plastics recycling across the country.
- Downy fabric softener will be available in refill packages. The consumer buys a large bottle of Downy, then follows up with smaller refill packages.
- Concentrated detergents. Superconcentrated detergents use far less packaging because the boxes and bottles they come in are smaller, but the ingredients are so concentrated a little bit goes a long way.
- Recycled paperboard. Most of P&G's paper boxes are made from recycled paper.

Colgate-Palmolive

Like many other companies, Colgate-Palmolive has instituted a corporate environmental policy. Their new policy states: "Colgate-Palmolive is committed to the protection of the environment in every area of the world in which we operate."

They are working to remove heavy metals from packaging, reduce the volume and weight of their packages, apply the plastics recycling code, as well as cleaning up their own manufacturing facilities and using energy wisely.

Bristol-Myers Squibb

In 1984, Bristol-Myers Squibb created its Corporate Environmental Committee to see what the company could do to clean up the environment. Since then, they have switched to using recycled paper in their Ban products and plan to do the same for all their paper-packaged goods.

Heinz

H. J. Heinz has changed the plastic fabrication of their squeezable ketchup bottles. The old design was made of too many layers of different plastics, making them difficult to recycle.

In 1991, Heinz will start using the new ENVIROPET ketchup bottle, which will be recyclable through existing PET (polyethylene terephthalate) systems. Working with other researchers, Heinz has produced a high-barrier PET container that is not only recyclable, but also fits into existing curbside collection and recycling networks for plastic PET soft-drink bottles.

SUPERMARKETS

Because they are the primary point of purchase for so many products, especially those that are heavily packaged, supermarkets are a flash point for environmental concerns.

Many supermarket chains have taken up the cause themselves:

- Bells Supermarkets in Buffalo, New York, and elsewhere, National Supermarkets in St. Louis and elsewhere and Big Bear in the San Diego area have developed, displayed and promoted lines of products such as unbleached coffee filters, tissue and toilet paper made of recycled paper, phosphate-free detergents and other items.
- Price Chopper, Edwards and Shop Rite supermarkets, all in the upstate New York and New England areas, as well as Giant in the Washington, D.C., area, have all started what amounts to grocery store recycling centers. Customers can bring their paper or plastic bags back to the store and drop them in recycling bins. Some of these stores as well as others now give 2-cent to 5-cent credit to people who bring their bags back to be reused. Some stores are accepting Styrofoam containers for recycling. Giant is even testing using compressed natural gas as a fuel in its vehicles to reduce air pollution.

Recycled toilet paper is everywhere. Tree Free, C.A.R.E., an acronym for Consumer Action to Save the Environment, EnviroCARE, Project GREEN and A.W.A.R.E. are just some of the brands of toilet paper, facial tissue, napkins and paper towels now being made of recycled paper. Even big names like Kimberly-Clark and Marcal are touting recycled paper products.

Recycled plastic bags are coming on strong. Consumers use 23 billion plastic grocery sacks each year. Mobil Chemical wants to recycle those bags and make them into new ones. They are already working with A&P, Kroger and Safeway. Plastic bags take up one-tenth of the room in the landfill as paper bags and you don't have to cut down trees to get plastic bags. To combat plastic's actions, paper bag manufacturers are starting to incorporate some recycled fibers in their brown bags.

RESTAURANTS

Restaurants, especially those that feature take-out service, have come under a lot of pressure from environmentalists and politicians for the amount of packaging they use, which ultimately ends up in our already overburdened landfills. Restaurants that use polystyrene foam food containers get even more heat because of the public's animosity toward polystyrene.

But Professor William Rathje, of the University of Arizona Garbage Project, has found that fast food packaging comprises only ⅓ of 1 percent of trash in a landfill. Furthermore, the polystyrene foam packages used in fast food restaurants are not made with ozone-depleting CFCs. Even though the public is misinformed, there is still a lot the restaurant industry can and should do to repair its public image.

Harris "Bud" Rusitzky, who recently stepped down as president of the National Restaurant Association, has some strong advice for restaurateurs concerning the environment.

"The environment will be as important in the 1990s as nutrition was in the 1980s," Rusitzky said in Chicago in June 1990. "Our customers will ask, and then insist, that we make a positive contribution to the ecology of the planet."

Rusitzky says customers will want restaurants to switch to reusable rather than disposable utensils and trays, cut down on the amount of packaging they use, even curtail the types of cleaning fluids they use.

The NRA and Rusitzky want restaurateurs to become proactive and to start doing the environmental things they can do before state and local governments start enforcing bans on them. He says restaurants can start using cloth or recycled paper napkins, print the menus on recycled paper. They can switch to reusable containers, look for ways to reduce their packaging requirements, even work with utilities to buy more energy-efficient cooking and refrigeration equipment. Do an energy audit on your building.

McDONALD'S AND THE ENVIRONMENT

Nobody has taken more heat for their polystyrene than McDonald's. Ironically, McDonald's switched from paper to plastic packaging in the mid-1970s because of their concern about cutting down so many trees.

To combat this heat, McDonald's launched its McRecycle USA campaign in April 1990. They have pledged to buy $100 million a year of recycled products, including roofing and insulation materials and playground equipment, to build and remodel its restaurants. Other items:

- McDonald's was the first company to ban CFCs from its polystyrene packaging in 1987.
- Four hundred fifty McDonald's restaurants in New England are separating polystyrene foam packages and recycling them.
- McDonald's spends $60 million a year on recycled paper products for its tray liners, paper napkins and other paper products.
- They have reduced the weight, thickness and con-

figuration of some of their packaging and boxes, reducing waste by 18 million pounds a year.

- McDonald's does not purchase beef from areas that have destroyed rain forest land to raise cattle.

But McDonald's isn't the only restaurant group that has begun to change its strategy to become an environmentally friendly food provider.

- Burger King is experimenting with shredding and composting its paper waste to reduce the amount of trash it sends to the landfill.
- Gorin's Homemade Ice Cream group in Atlanta has started to recycle its plastic ice cream barrels, aluminum cans, cardboard boxes and paper products.
- The Hyatt Regency Hotel in Chicago has established an on-site recycling center to reduce its 7.2 million pounds of annual trash. They are recycling cardboard, plastic, glass and aluminum cans.
- Hard Rock Cafes in many locations have started recycling glass, cardboard, paper and aluminum. They use only degradable or recyclable serving containers and they donate money to environmental causes.

TEN COMMANDMENTS FOR ENVIRONMENTAL MARKETING

1. Avoid greenwashing.

As in whitewashing. Don't try to cover up an environmental problem you have by making a donation to an environmental group or planting a few trees. Approach your problem head-on and either change it or justify your actions to your critics.

2. Don't jump on the green bandwagon.

Don't be hasty in latching on to the latest environmental panacea. Witness the chemical companies that got their heads bumped when they started selling biodegradable plastics. Procter & Gamble held out against degradable diapers and they have been proven correct.

3. Think cradle to grave.

Whenever you are designing a new product, package or activity, analyze the life cycle of that item, from the oil in the ground or trees in the forest all the way to its end use in the landfill or incinerator. Think through all the possible environmental ramifications of every move.

4. Look in-house first.

Before you redesign your marketing, analyze what you are already doing to see what environmental benefits you have. Kellogg's just received an award from an environmental group for using a high percentage of recycled paper in their cartons. They have been doing it for most of the twentieth century and nobody made a fuss before. Look around, you may have an exceptionally clean manufacturing process, or a good ride-sharing program or nonhazardous cleaning products in your building.

5. Think global—act local.

Don't focus your attention on rain forests in Brazil. There are plenty of environmental problems in the United States, Canada and Mexico to keep us all busy. Support your local environmental groups who want to clean up the Long Island Sound or save the Illinois prairies before you contribute to big national and international environmental organizations that have greater resources than your local groups.

6. Get to know your critics.

Seek input from your so-called opponents. Listen to them rather than trying to co-opt them. They have spent a lot of time analyzing your environmental problems and they may have good information you can use to find a solution. In July 1990, McDonald's entered into an agreement with the Environmental Defense Fund to study ways for the restaurant chain to cut down on the amount and type of garbage they produce in their operations.

7. Make sense with tie-in promotions.

There is nothing wrong per se with tie-in promotions. Donating a nickel to a tree-planting program for every bottle of beer you sell, or lending your corporate name and support for a worthy environmental cause, are good things to do.

A tie-in promotion is a way for you to let the public know you really care about the cause or program you are supporting. Make sure your tie-in is for the long run and you don't appear to be jumping from one cause to the next. Make a logical connection between what your company makes or sells and your tie-in. For example, the makers of Glad bags support beach cleanup efforts.

Other ideas:

- If you own a restaurant that creates a lot of litter, help sponsor a recycling center.
- If you are an architectural or engineering firm, sponsor a solar design contest.

Finally, remember, the public will appreciate you more if you roll up your sleeves and tackle difficult problems like air pollution caused by commuter traffic, or start a vigorous office paper recycling plan. People respect hard work and diligence.

8. Be humble.

Whenever you do have success in solving an environmental problem or accomplishing one of your environmental goals, be gracious about informing your customers, your stockholders and the public. If you brag about it, someone might wonder, "If it was so easy to do, why didn't you do it twenty years ago?"

9. The green edge.

Realize that environmentally responsible packaging, products and processes can give you a distinct advantage over your competitor. Everybody agrees that the environment will be the top story and concern of the 1990s. The sooner you clean up your environmental act, the bigger jump you get in the marketplace.

10. Environmental ethics equal profits.

It is very possible that sound environmental practices will mean profits for you and your stockholders. But the most important thing you should remember when you analyze your company's environmental practices is that you should do it because it is the right thing to do. Let ethics and responsibility be your guide. Let the voices of children who care about silly things like dolphins, whales and squirrels ring in your ears. Remember, a cleaner environment is a corporate and individual legacy you can leave behind, a legacy remembered long after you are gone.

7

Environmental Investing, or The Color Green

> The business of cleaning up the environment and keeping it clean will spawn the greatest growth industry the world has seen since the flowering of the military-industrial complex following World War II.
>
> —*USA Today,* July 7, 1989

Green has always been the most popular color on Wall Street and in other financial centers around the United States. But now, with an environmental president in the White House and the public thinking more and more about the environment and what it will cost to clean it up, the color green has taken on new meaning in the investment community.

Wall Street is rushing to create new investment funds and venture cash pools to be in on the ground floor of what many analysts and experts expect to be a major economic growth area in the 1990s.

At least six new funds were created in the six months leading up to Earth Day 1990. Many of them were designed to provide investment avenues for the solid and hazardous waste handling firms, recyclers, toxic waste cleanup specialists, air pollution emission abatement companies and others.

But several of the funds were designed for investors who do not want their money going into the coffers of known polluters. The funds and their advisers screen the companies to be sure they don't contribute to environmental degradation by their products or practices.

This chapter will take a look at the new investment funds, both the market-driven cleanup funds as well as the clear-conscience socially responsible funds.

We will also discuss the pressure investors, many of them large institutional investors, who are putting pressure on companies through stockholder proxy initiatives as well as promoting environmental charter resolutions such as the *Valdez* Principles.

There are also many other financial vehicles that can affect the environment, such as environmental checking accounts, banking services, IRAs, even long-distance services.

Finally, we will take a look at new business: what entrepreneurs are doing already, what we as a business and governmental community need to do to encourage them.

DISCLAIMER: This book does not make any specific recommendations for readers to use as investment strategy. We are not suggesting you invest in any funds or companies listed here. We are only offering you a survey of the current situation.

All investments carry certain risks and we suggest you work with qualified professionals when making decisions. The emerging environmental cleanup growth stocks are considered risky by many investment professionals. Furthermore, socially conscious investing is a form of activist investing that can be used to pressure companies on moral and environmental issues. We are not taking sides, only presenting information.

INVESTING IN ENVIRONMENTAL CLEANUP

Many business people are saying that the business of cleaning up the environment will be the growth issue of the 1990s. One hundred billion dollars were spent on environmental cleanup in 1989. New initiatives to reduce acid rain and remove asbestos and lead paint could add even more.

Clean Air—The new Clean Air Act will require lowered toxic and fossil fuel emissions from cars, power plants and factories. New technologies, better systems and cleaner-burning fuels will have to be developed. Nonpolluting automobiles and solar power are returning to the marketplace.

Solid Waste and Recycling—Already ten states and dozens more counties and municipalities have started mandating source separation and recycling. There is a rush to invent new shredders, chippers and grinders to transform materials such as tires, newspapers and plastic bottles into useful products. New trucks to handle the separated refuse and recovery facilities have to be built.

Hazardous Waste—Congress is scheduled to take up the Resource Conservation and Recovery Act and Superfund authorization in 1990, which will assuredly give new emphasis to companies and technologies that can clean up the toxic dump sites across the USA.

Now, brokerage firms have been tracking and investing in cleanup firms for years and you don't really need any special system to make investments. But there has also been a trend on Wall Street recently to create targeted fund groups to appeal to investors who want their money to go in certain directions. These funds are also a good way to attract investor interest.

The main school of thought on cleanup investments is that there will be a lot of demand for these companies' goods and services in the next decade and that they will be good money makers for people who invest there.

Another school of thought says that these cleanup companies are major players in an area that is critical to the long-term environmental health of our nation and our planet. By investing in the ones that are considered the most conscientious, such as AT&T, which is phasing out using CFCs in their telephone equipment, you are encouraging AT&T to keep up the good work and suggesting to other players that they do the same.

THE DOWNSIDE: The downside on investing in cleanup companies is that environmental cleanup is risky for two reasons. One, they are prone to lawsuits because of the problems inherent in handling toxic waste and emissions, and it can be very difficult sometimes to remain within EPA guidelines on emissions.

Second, they are risky because the market for environmental goods and services is volatile. One day you invest in companies making biodegradable plastic bags and the next day those same bags are being blasted by environmentalists at a news conference. The bottom line is that you should follow all prudent rules of investing in risky growth stocks that have enormous potential.

Here is a partial listing of some of the cleanup funds. It is meant to indicate the sorts of funds available but does not include them all. This book does not make any recommendations. Information is compiled from publicly available prospectuses, newsletters and other sources.

Freedom Environmental Fund, One Beacon Street, Boston, MA 02108, 617-523-3170, 800-225-6258 nationwide, 800-392-6037 in Massachusetts.

Part of the John Hancock group, began in August 1989. Invests primarily in solid waste handlers like Waste Management, Inc., water treatment and equipment, natural gas, air pollution control and hazardous waste handlers.

Fidelity Select Environmental Services, The Fidelity Building, 82 Devonshire Street, Boston, MA 02109, 800-544-8888.

One of Fidelity's thirty-six portfolios, started in June 1989. Invests in companies "engaged in the research, development, manufacture and distribution of products, processes and services related to waste management and pollution control." Fidelity also has a select portfolio on energy that looks at geothermal, wind and solar energy alternatives.

Environmental Awareness Fund, 1016 W. 8th Avenue, King of Prussia, PA 19406, 215-337-8422, 800-523-2044.

A part of the SFT family of funds, started November 1988. Finished seventh in the *Wall Street Journal,* March 30, 1990, Mutual Fund Scoreboard of Top 15 Performers. Features fifty-four companies involved in air pollution control, waste management, resource recovery, alternate energy and others. Some OTC, some American and NYSE stocks.

Other funds include Alliance Global Environment, 800-227-4618; Environmental Appreciation Fund, 800-421-9900; Kemper Environmental Services, 800-621-1048; Oppenheimer Global Environment, 800-525-7048; and Merrill Lynch Environmental Unit Investment Trust.

There is also a green investing newsletter, *Wall Street Green Review,* Environmental Investment Strategies, 24861 Alicia Parkway, Suite C-293, Laguna Hills, CA 92653. Subscription is $96 a year, edited by David Lawrence Brown.

ENVIRONMENTALLY CONSCIOUS INVESTING

Environmentally conscious investing is designed for people and institutions who put as high a premium on the environmental purity of the companies they invest in as they do on dividends. In addition, these investors feel strongly that a company's record on environmental cleanup will have a significant effect on the company's bottom line in the next decade and beyond. Finally, these investors feel that by using their investment dollars, they can influence a company's policy on environmental matters.

Typically, these investors are people with trust funds. But institutions like churches and some large public assets like pension funds are entering this market or putting "green" restrictions on their investments. There are also several funds opening up to serve these investors.

This concept of environmentally aware investing is an outgrowth of the anti–South African apartheid disinvestment strategies of the 1980s. These strategists have taken partial credit for Nelson Mandela's welcome to the United States last summer.

"Typically, these funds will screen a company to see if it has any EPA violations, lawsuits from environmental groups, if its products create an environmental risk, and if it owns any Superfund hazardous waste sites," says Scott Fenn, director of the energy and environment programs of the Investor Responsibility Research Center, in Washington, D.C.

Fenn's group will be researching *Fortune* 500 blue chip companies for environmental problems as well as profitability and will have its first reports available for sale in early 1991. They traditionally have serviced large pension funds with research but are looking to branch out to other investors.

These funds typically list their board of directors or board of advisers to show investors how committed they are to environmentally correct investing.

Here is a partial and informational listing of these funds, compiled from prospectuses and other sources. Again, this book does not make any recommendations or take any sides.

Working Assets, 230 California Street, San Francisco, CA 94111, 415-989-3200, 800-533-3863.

Began operation in 1983, now with $210 million in assets. Working Assets screens companies for compliance with EPA and OSHA regulations, if they have a record with EPA or the Justice Department of violating regulations, if they supply nuclear equipment or own nuclear power plants. They encourage firms involved in alternate energy or energy conservation.

Calvert Social Investment Fund, 1700 Pennsylvania Avenue, NW, Washington, D.C. 20006, 301-951-4240, 800-638-6731.

Started in 1982, Calvert screens companies on environmental issues as well as restricting investments in those who do business in South Africa, produce nuclear weapons or are involved in producing nuclear weapons or are involved in producing nuclear energy.

Other socially and environmentally responsible money funds include New Alternatives Fund, Great Neck, New York (516-466-0808), mainly solar and alternative energy investments, Progressive Securities, Portland and Eugene, Oregon (503-224-7828), Pax World Fund, Portsmouth, New Hampshire (603-431-8022), The Parnassus Fund, San Francisco (415-362-3505).

In addition to conventional investments, there are banking, IRA, long-distance telephone and check-writing services that put your money to work for environmental causes.

Socially Responsible Banking Fund, Vermont National Bank, P.O. Box 84, Brattleboro, VT 05302-9987, 802-257-7151, 800-544-7108.

This is an individual investor savings fund in CDs, IRAs, money market, savings and checking with interest that creates a fund that loans money to targeted areas. Fifteen percent of the loan money goes to environmental/conservation projects for small business development in recycling, alternate energy and land conservation.

Working Assets Earth Day IRA and Long Distance, 800-877-2100.

If you had opened an IRA with Working Assets up until April 16, 1990, a donation of $1 for every $1,000 you invested would have been donated to Earth Day 1990. This IRA may be opened again.

Working Assets Long Distance, in cooperation with U.S. Sprint, will give 1 percent of your charges to worthy environmental groups such as the Rainforest Action Network or the Environmental Defense Fund.

Environmental MasterCard or Visa.

Full-service bank cards that have linked up with environmental groups. Use one of these cards and give a donation to one of these groups:

- Chesapeake Bay Foundation, 800-847-7378, 301-268-8816.
- Defenders of Wildlife, 800-548-4783, 202-659-9510.
- Environmental Defense Fund, 800-833-3010, 212-505-2375.
- National Wildlife Federation, 202-797-6800, 800-847-7378.

SHAREHOLDER ACTIVISM AND THE *VALDEZ* PRINCIPLES

Shareholder activism around environmental issues is a major new trend in investing. For years, groups have bought shares in a company and then attended the annual meeting to raise hell over particular issues.

But the new development now is that the shareholder activists' initiatives are getting on the proxy and forcing votes on environmental policy at annual meetings. Exxon faced six shareholder proposals at their annual meeting in spring 1990.

Further adding fire to these proxy fights is the new trend in environmental screening by large shareholding groups like churches and city- and state-owned pension funds. Cities and states are proposing legislation that would apply environmental guidelines to their investment strategies. When these institutional investors vote at the annual meeting and in the market, companies will be listening.

An example of this new trend is the Coalition for Environmentally Responsible Economies, or CERES, a project of the Boston-based Social Investment Forum.

The CERES project is encouraging all companies and corporations to sign and abide by these principles and they are encouraging shareholders to force the issue if they won't.

As of the summer of 1990, no major companies had adopted the *Valdez* Principles. Companies should expect these principles or ones like them to be a regular part of the annual meeting each spring.

THE *VALDEZ* PRINCIPLES

Introduction

By adopting these principles, we publicly affirm our belief that corporations and their shareholders have direct responsibility for the environment. We believe that corporations must conduct their business as responsible stewards of the environment and seek profits only in a manner that leaves the Earth healthy and safe. We believe that corporations must not compromise the ability of future generations to sustain their needs.

We recognize this to be a long-term commitment to update our practices continually in light of advances in technology and new understandings in health and environmental science. We intend to make consistent, measurable progress in implementing these principles and to apply them wherever we operate throughout the world.

1. Protection of the Biosphere

We will minimize and strive to eliminate the release of any pollutant that may cause environmental damage to the air, water or earth or its inhabitants. We will safeguard habitats in rivers, lakes, wetlands, coastal zones and oceans and will minimize contributing to the greenhouse effect, depletion of the ozone layer, acid rain and smog.

2. Sustainable Use of Natural Resources

We will make sustainable use of renewable natural resources, such as water, soils and forests. We will conserve nonrenewable natural resources through efficient use and careful planning. We will protect wildlife habitat, open spaces and wilderness, while preserving biodiversity.

3. Reduction and Disposal of Waste

We will minimize the creation of waste, especially hazardous waste, and wherever possible recycle materials. We will dispose of all wastes through safe and responsible methods.

4. Wise Use of Energy

We will make every effort to use environmentally safe and sustainable energy sources to meet our needs. We will invest in improved energy and conservation in our operations. We will maximize the energy efficiency of products we produce or sell.

5. Risk Reduction

We will minimize the environmental, health and safety risks to our employees and the communities in which we operate by employing safe technologies and operating procedures and by being constantly prepared for emergencies.

6. Marketing of Safe Products and Services

We will sell products or services that minimize adverse environmental impacts and that are safe as consumers commonly use them. We will inform consumers of the environmental impacts of our products or services.

7. Damage Compensation

We will take responsibility for any harm we cause to the environment by making every effort to fully restore the environment and to compensate those persons who are adversely affected.

8. Disclosure

We will disclose to our employees and to the public incidents relating to our operations that cause environmental harm or pose health or safety hazards. We will disclose potential environmental, health or safety hazards posed by our operations, and we will not take any action against employees who report any condition that creates a danger to the environment or poses health and safety hazards.

9. Environmental Directors and Managers

At least one member of the Board of Directors will be a person qualified to represent environmental interests. We will commit management resources to implement these Principles, including the funding of an office of vice president for environmental affairs or an equivalent executive position, reporting directly to the CEO, to monitor and report upon our implementation efforts.

10. Assessment and Annual Audit

We will conduct and make public an annual self-evaluation of our progress in implementing these Principles and in complying with all applicable laws and regulations throughout our worldwide operations. We will work toward the timely creation of independent environmental audit procedures which we will complete annually and make available to the public.

For more information on the *Valdez* Principles, contact CERES at 711 Atlantic Avenue, Boston, MA 02111, 617-451-3661 or 617-451-3252.

NEW BUSINESS

According to George G. Montgomery, Jr., an investment banker with Hambrecht & Quist in New York, the major stumbling block in the creation of new business in the environmental field is the lack of entrepreneurs.

"There's plenty of capital in the marketplace," says Montgomery, who is also a board member of the Environmental Defense Fund. "What the environment needs are people who are really good at building a business.

"So far we just have not been attracting the talented kinds of people to pollution-related business like we did in the computer or medical equipment fields."

Montgomery says the reason for this is that no entrepreneur in the environmental field has hit a financial home run the way they did at Microsoft or Apple. There are no Steve Jobses in the environmental business to act as role models.

Montgomery says the solution is to create a batch of marketplace incentives to encourage entrepreneurs. He suggests putting a tax on gasoline to reflect its true environmental costs to encourage alternative fuel or transportation system development. Or a tax on existing energy sources to make solar more profitable.

As Montgomery says, there is plenty of money available for new environmental business. The *Wall Street Journal* reported that over $100 million has been invested in enviro-firms in the last two years. The National Association of Manufacturers estimates that business will spend $49 billion a year on environmental improvements.

THE BOTTOM LINE

Look for growth as an investor or entrepreneur in:

- Hazardous and solid waste cleanup and handling.
- Biotechnology to remedy spills and emissions.
- Engineering firms that are designing new systems.
- Consultants and public relations specialists who are advising companies on creating greener companies.
- Water and air treatment specialists.

About the Authors

Laurence Sombke, the author of *The Solution to Pollution: 101 Things You Can Do to Clean Up Your Environment,* is a New York–based writer who has written for *USA Today, USA Weekend* (for which he recently wrote a special supplement entitled "Clean Up Your Own Backyard"), *Esquire, New York* magazine, *Family Circle* and other publications. Sombke has produced and announced "Energy Issues," a program heard on radio stations throughout the Midwest, and also has been a newswriter for ABC Radio News in New York. He has an M.A. in journalism.

Terry M. Robertson, a chemical engineer, is executive vice president of STRATCO, Inc., an international chemical engineering firm. Elliot M. Kaplan, a lawyer, has held top-level management posts in the automobile and airline industries. Both Robertson and Kaplan live in the Kansas City area.

Additional copies of *The Solution to Pollution in the Workplace* may be ordered by sending a check for $9.95 (please add the following for postage and handling: $1.50 for the first copy, $.50 for each added copy) to:

MasterMedia Limited
16 East 72nd Street
New York, NY 10021
(212) 260-5600
(800) 334-8232

Laurence Sombke is available for speeches and workshops. Please contact MasterMedia's Speakers' Bureau for availability and fee arrangements. Call Tony Colao at (201) 359-1612.

For interviews, personal appearances, spokesperson tours, product endorsements and consulting, contact Laurence Sombke directly at P.O. Box 354, Claverack, NY 12513, (518) 851-5521.